道路標識・標示一覧表 (List of Traffic Signs and Markings)

規制標示 Regulatory Markings

回転禁止 No U-Turn	追越しのための右側部分はみ出し通行禁止 No Overtaking on the Right Side Over the Marked Area		
図示の 8-20 は，車両の転回を禁止する時間が 8 時から 20 時までであることを示す Diagram 8-20 indicates that U-turn of vehicles is prohibited from 8:00 to 20:00.	AおよびBの部分の右側部分はみ出し追越し禁止 No Overtaking on the Right Side Over the Marked Areas A and B	AおよびBの部分の右側部分はみ出し追越し禁止 No Overtaking on the Right Side Over the Marked Areas A and B	Bの部分からAの部分へのはみ出し追越し禁止 No Overtaking Over the Marked Area from B to A
進路変更禁止 No Lane Change		駐停車禁止 No Stopping or Parking	駐車禁止 No Parking
Aの車両通行帯を通行する車両がBを通行することおよび，Bの車両通行帯を通行する車両がAを通行することを禁止 Vehicles traveling on lane A are prohibited from entering lane B, and vehicles traveling on lane B are prohibited from entering lane A.	Bの車両通行帯を通行する車両が，Aの車両通行帯を通行することを禁止 Vehicles traveling on lane B are prohibited from entering lane A.		

道路標識・標示一覧表 （List of Traffic Signs and Markings）

最高速度 Maximum Speed Limit	立入り禁止部分 No Entry Area	停止禁止部分 No Stopping Area	路側帯 Side Strip
30			路側帯　車道

車両通行帯 Vehicle Lane			
ペイントなどによるとき 高速自動車国道の本線車道以外の道路の区間に設けられる車両通行帯 When Marked or Painted: Lanes on sections of roads other than the main lanes of the expressway.		道路びょうなどによるとき When indicated by road studs, etc:	高速自動車国道の本線車道に設けられる車両通行帯 Lanes provided on the main lanes of the expressway.

駐停車禁止路側帯 No Parking Zone on the Side Strips	歩行者用路側帯 Pedestrian Sidewalk	優先本線車道 Priority Main Lane	車両通行区分 Traffic Distribution
路側帯　車道	路側帯　車道	本線（優先）・本線	自動車・二輪を除く／二輪・軽車両
車の駐車と停車が禁止されている路側帯であることを示す The marking indicates a side strips where parking and stopping of vehicles are prohibited.	車の駐停車，軽車両の通行が禁止されている路側帯であることを示す The marking indicates a side strip where parking and stopping of vehicles, as well as the passage of light road vehicles, are prohibited.	この標示がある本線車道と合流する前方の本線車道が，優先道路であることの指定 The main lane ahead, merging with the main lane displaying this marking, is a priority road.	図示の文字は，通行区分を指定された車両通行帯と車の種類を示す The characters specify the vehicle lane with designated traffic categories and indicate the types of vehicles.

2　道路標識・標示一覧表　(List of Traffic Signs and Markings)

特定の種類の車両の通行区分 Traffic Distribution for Specific Types of Vehicles	けん引自動車の高速自動車国道通行区分 Expressway Traffic Distribution for Towing Vehicles	専用通行帯 Exclusive Lane	路線バス等優先通行帯 Bus Priority Lane

進行方向別通行区分 Direction-Specific Traffic Distribution	右左折の方法 Procedures for Right and Left Turns		

けん引自動車の自動車専用道路第一通行帯通行指定区間 Traffic Distribution for the First Lane of Exclusive Roads for Towing Vehicles	平行駐車 Parallel Parking	直角駐車 Perpendicular Parking	斜め駐車 Diagonal Parking
	 1台　2台以上 One Vehicle Vehicles		

道路標識・標示一覧表　(List of Traffic Signs and Markings)

普通自転車歩道通行可 Bicycles Allowed on Sidewalk	普通自転車の歩道通行部分 Section for Bicycles on The Sidewalk	普通自転車の交差点進入禁止 No Entry for Bicycles at Intersection	終わり End of The Traffic Regulations
	普通自転車が歩道を通行することができることと，その場合に通行しなければならない部分の指定 The zone indicates that regular bicycles are allowed to use the sidewalk and specifies the section they must use when passing.	普通自転車が，この標示をこえて交差点に進入するのを禁止することを示す The marking prohibits regular bicycles from entering the Intersection beyond this marking.	規制標示が表示する交通規制の区間の終わりであることを示す The markings indicates the end of the traffic regulation section.

指示標示　Instructional Sign

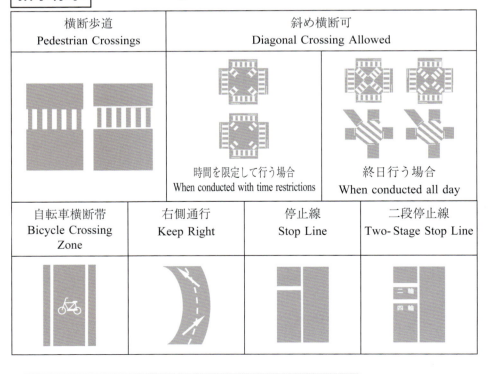

4　道路標識・標示一覧表　(List of Traffic Signs and Markings)

進行方向
Direction of Travel

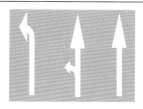

車線境界線
Lane Boundary Line

	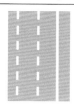	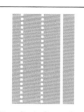
ペイントなどによるとき When indicated by paint, etc.	ペイントなどによるとき When indicated by paint, etc.	道路びょうなどによるとき When indicated by markings by road studs, etc.

中央線
Center Line

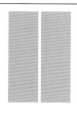		

1. 道路の右側にはみ出して通行してはならないことを特に示す必要がある道路に設ける場合
1. When driving on roads where it is necessary to specifically indicate the prohibition of protruding to the right side of the road.

2. 1以外の場所に設ける場合
2. If provided in locations other than in 1.

(1) ペイントなどによるとき
(1) When indicated by paint, etc.

(2) 道路びょうなどによるとき
(2) When indicated by markings by road studs, etc.

道路標識・標示一覧表 (List of Traffic Signs and Markings)

3. 道路の中央以外の部分を道路の中央として指定する場合 3. If designating a section other than the center of the road as the center of the road.		4. 1と3の(1)の場合で，特に必要があるとき 4. In the case of (1) in both 1 and 3, when necessary.
(1) 常時指定するとき (1) When designated at all times.	(2) 日や時間を限って指定するとき (2) When designated for specific days or times.	

安全地帯 Safety Zone	安全地帯または路上障害物に接近 Approaching Safety Zone or Road Obstacle		導流帯 Guide Strip
	片道にさける場合 When passing on one side	両側にさける場合 When passing on both sides	車の通行を安全で円滑に誘導するため，車が通らないようにしている道路の部分であることを示す The marking indicates a road section designed to guide the safe and smooth passage of vehicles by preventing them from entering.

6　道路標識・標示一覧表　(List of Traffic Signs and Markings)

路面電車停留場 Streetcar Stop	横断歩道または自転車横断帯あり Pedestrian or Bicycle Crossing Ahead	前方優先道路 Priority Road Ahead

規則標識　Regulatory Sign

通行止め Road Closed	二輪の自動車原動機付自転車通行止め No Entry for Two-wheeled Vehicles and Motorized Bicycles	車両横断禁止 No Right Turn Crossing Ahead	駐車禁止 No Parking	最高速度 Maximum Speed Limit
車両通行止め No Entry for Vehicles	大型自動二輪車及び普通自動二輪車二人乗り通行禁止 Prohibition of Riding with Two Persons on Large-size Motorcycles and Regular Motorcycles	転回禁止 No U-Turn	駐車余地 Parking Space	特定の種類の車両の最高速度 Maximum Speed for Specific Types of Vehicles
車両進入禁止 No Entry for Vehicles	自動車以外の軽車両通行止め No Entry for Light Road Vehicles Except Automobiles	追越しのための右側部分はみ出し通行禁止 No Passing on the Right-Hand Part of the Road for Overtaking	時間制限駐車区間 Time-limited Parking Zone	最低速度 Minimum Speed Limit

道路標識・標示一覧表　(List of Traffic Signs and Markings)　7

二輪の自動車以外の自動車通行止め No Entry for All Vehicles Except Two-Wheel Vehicles	自転車通行止め No Entry for Bicycles	追越し禁止 No Overtaking 追越し禁止	危険物積載車両通行止め No Vehicles Carrying Dangerous Substances 危険物	自動車専用 Motor Vehicles Only
大型貨物自動車等通行止め No Entry for Large-size Trucks, etc.	車両（組合せ）通行止め No Entry for Vehicles Shown in The Sign	駐停車禁止 No Stopping or Parking 8 - 20	重量制限 Maximum Weight Limit 5.5 t	自転車専用 Bicycles Only
特定の最大積載量以上の貨物自動車等通行止め No Entry for Freight Vehicles, etc. above a Specified Maximum Loading Capacity 積3t	指定方向外進行禁止 Proceed Only in the Designated Direction	高さ制限 Maximum Height Limit 3.3m	自転車及び歩行者専用 Bicycles and Pedestrians Only
大型乗用自動車等通行止め No Entry for Large Passenger Vehicles, etc.		最大幅 Maximum Width Limit 2.2m	歩行者専用 Pedestrian Paths

8　道路標識・標示一覧表　(List of Traffic Signs and Markings)

一方通行 One-way	路線バス等優先通行帯 Priority Lane for Route Buses	環状の交差点における右回り通行 Clockwise Circulation at Roundabout	前方優先道路 Priority Road Ahead
自転車一方通行 One-way for Bicycles	普通自転車専用通行帯 Bicycle-Only Lane	平行駐車 Parallel Parking	一時停止 Stop

道路標識・標示一覧表 （List of Traffic Signs and Markings）

車両通行区分 Vehicle Lane Distriution	進行方向別通行区分 Lane Direction Distribution	直角駐車 Perpendicular Parking	専用通行帯 Exclusive Lane
特定の種類の車両の通行区分 Lane Distribution for Specific Vehicles		斜め駐車 Diagonal Parking	歩行者通行止め No Entry for Pedestrians
けん引自動車の高速自動車国道通行区分 Lane Distribution for Towing Vehicles		警笛鳴らせ Sound Horn	歩行者横断禁止 No Crossing
けん引自動車の自動車専用道路第一通行帯通行指定区間 The Designated Lane for Towing Vehicles on Motorway	原動機付自転車の右折方法（二段階） Two-stage Right Turn for Motorized Bicycles	警笛区間 Sounding Horn Zone 	原動機付自転車の右折方法（小回り） The Direct Right Turn for Motorized Bicycles 徐行 Reduced Speed 

10　道路標識・標示一覧表　(List of Traffic Signs and Markings)

指示標識	Direction Signs		
並進可 (Allowed to) Go Alongside	中央線 Center Line	高齢運転者等標章自動車停車可 Stopping Permitted for Elderly Drivers, etc., with the Special Identification Mark	自転車横断帯 Bicycle Crossing
軌道敷内通行可 Driving on Trucks Permitted	停止線 Stop Line	停車可 Stopping Permitted	横断歩道・自転車横断帯 Pedestrian Crossing / Bicycle Crossing
高齢運転者等標章自動車駐車可 Stopping Permitted for Elderly Drivers, etc., with the Special Identification Mark	横断歩道 Pedestrian Crossing	優先道路 Priority Road	安全地帯 Safety Zone
駐車可 Parking Permitted		規制予告標示板 Regulatory Notice Board	

道路標識・標示一覧表 （List of Traffic Signs and Markings） 11

補助標識　Supplementary sign

距離・区域 Distance and Area	日・時間 Date and Time	車両の種類 Type of Vehicle	駐車余地 Parking Space
この先100m ここから50m 市内全域	日曜・休日を除く 8-20	大　貨 原付を除く 積3t	駐車余地6m

始まり Start of Traffic Regulation	区間内・区域内 Zone or Section of Traffic Regulations	終わり End of Traffic Regulations	追越し禁止 No Overtaking
ここから 区域 ここから	区域内	ここまで 区域 ここまで	追越し禁止

通学路 School Road	踏切注意 Beware of Railroad Crossing	横風注意 Beware of Crosswind	動物注意 Beware of Animals
通学路	踏切注意	横風注意	動物注意

注意 Caution	注意事項 Precautions	規制理由 Reason for Traffic Regulation	方向 Direction
注　意	路肩弱し 安全速度 30	騒音防止区間 歩行者横断多し 対向車多し	

地名 Place Name	前方優先道路 Priority Road Ahead	始点 Starting Point	終点 Terminus
小諸市 本町	前方優先道路	始　点	終　点

標示板など Signs, etc.

信号に関わらず左折可能であることを示す標示板 A Sign Indicating That Left Turns Are Permitted Regardless of the Traffic Signals 	特定の交通に対する信号機の標示板 Signal Indicator Board for Specific Traffic 	仮免許練習標識 Provisional License Practice
車輪止め装置取付け区間であることを示す表示板 A Sign Indicating a Section Where Wheel Stop Devices Are Installed 始まり ?? 区間内 ??? 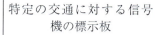 終わり ?????	指定消防水利の標識 Designated Fire Hydrant 	代行運転自動車標識 Designated Driver Service 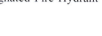
	初心運転者標示機 Novice driver Sign 	パーキング・チケット発給設備があることを示す表示板 Signboards Indicating Machines for Issuing Parking Ticket
時間制限駐車区間があることを示す表示板 Time-limited Parking Zone 	身体障害者標識 Handicapped Mark 	高齢運転者標識 Senior Driver Sign 聴覚障害者標識 Hearing-impaired Driver Mark

道路標識・標示一覧表 (List of Traffic Signs and Markings)

案内標識　Guide Sign

入口の予告 Advance Notice of Entrance	非常電話 Emergency Phone	方面と車線 Direction and Lane
名神高速 MEISHIN EXPWY 入口 150m	非常電話	大阪 Osaka 本線 THRU TRAFFIC
登坂車線 Climbing Lane	傾斜路 Slope	乗合自動車停留所 Bus Stop
登坂車線 SLOWER TRAFFIC 登坂車線 SLOWER TRAFFIC		バスのりば バスのりば
方面，車線と出口の予告 Advance Notice of Direction, Lane, and Exit	方面と出口 Direction and Exit	駐車場 Parking Lot
京都 宇治 Kyoto Uji 5B 出口 1km EXIT 江戸橋 Edobashi 303 出口 400m EXIT	横浜 町田 Yokohama Machida 4 出口 EXIT 西神田 Nishikanda 501 出口 EXIT	P P
待避所 Shelter	サービスエリアの予告 Advance Notice of Service Area	入口の方向 Direction of Entrance
待避所	P 1km 富士川 Fujigawa P 1km 中井 Nakai	東名高速 首都高速 TOMEI EXPWY SHUTO EXPWY 空港 新宿 Airport Shinjuku
方面と方向の予告 Advance Notice of Direction and Destination	方面と距離 Direction and Distance	方面，方向と道路の通称名 Direction, Route and Road Nickname
日本橋 Nihonbashi 上馬 大森 Kamiuma Oomori 300m	日本橋 10km Nihonbashi 日比谷 7km Hibiya 4 横浜 11km Yokohama 5 厚木 26km Atsugi 静岡 153km Shizuoka	市ヶ谷 Ichigaya 池袋 渋谷 Ikebukuro Shibuya 明治通り 300m

14 道路標識・標示一覧表　(List of Traffic Signs and Markings)

道路標識・標示一覧表 (List of Traffic Signs and Markings) 15

合流交通あり Road Entry Left	車線数減少 Lane Reduction	幅員減少 Road Narrows	上り急こう配あり Steep Upward Slope
下り急こう配あり Steep Downward Slope	道路工事中 Road Works	動物が飛び出すおそれあり Animals Crossing	その他の危険 Other Hazards
Y形道路交差点あり Y- intersection	踏切あり Railroad Crossing	学校，幼稚園，保育所などあり School, Kindergarten, Nursery, etc.	路面凹凸あり Hump or Dip
二方向交通 Two- way Traffic	横風注意 Crosswind		

16 道路標識・標示一覧表 （List of Traffic Signs and Markings）

Explained in English
★ 英語で解説

原付免許
Motorized Bicycle License

めざせ 一発合格!
Let's Pass on the first try

監修　株式会社知財事業研究所
大賀信幸・山本英彦・柳井正彦【編著】

弘文社

はじめに

　本書は，日本に滞在している外国人の方々が，特に，原動機付自転車（原付バイク）の免許取得に役立てていただけるよう執筆されました。

　各国では，バイクの運転に関してさまざまな規制が適用されています。母国では，運転免許の規制が緩やかだったり，罰則がそれほど厳しくなかったりすることもあるかもしれません。しかし，日本には，日本独自の原動機付自転車のルールがあります。

　たとえ母国でバイクの免許を持っていたとしても，日本国内で有効な免許を取得または切り替えをしていなければ，日本では無免許運転と見なされます。そのため，日本でバイクに乗る際は，日本の法律に従い，必ず『原付免許』を取得することが求められます。

　『原付免許』の試験は，しっかりと準備すれば合格が可能な試験です。しかし，勉強せずに合格できるほど簡単ではありません。合格するためには，試験の出題傾向を理解し，模擬試験を通じて十分な対策を行うことが重要です。

　試験問題は，日本語で出題されることが一般的ですが，日本語が第一言語でない皆さんにとっては，言い回しや表現が普段の日本語よりも難しく感じられるかもしれません。また，日本の交通ルールやマナーに関する「引っかけ問題」も多く含まれているため，慎重な対策が必要です。

　本書は，こうした「引っかけ問題」への対策や，日本語の試験問題に慣れるための手助けとなることを目指しています。日本での安全な運転と，スムーズな生活のために，本書を活用して，ぜひ一発で『原付免許』を取得してください。

株式会社知財事業研究所

取締役　山本　英彦

Introduction

This book has been written to help foreign residents in Japan, especially those seeking to obtain a license for a motorized bicycle (moped license).

Each country has different regulations regarding motorcycle driving. In some countries, the regulations for driver's licenses may be more lenient, or penalties may not be as strict. However, Japan has its own unique rules for mopeds.

Even if you hold a motorcycle license in your home country, unless you obtain or convert it to a valid license within Japan, you will be considered driving without a license. Therefore, when riding a motorcycle in Japan, you are required to obtain a moped license in accordance with Japanese law.

The moped license exam is a test that can be passed with proper preparation. However, it is not so simple that you can pass without studying. To pass, it is crucial to understand the exam trends and take sufficient mock tests to prepare thoroughly.

The exam questions are typically written in Japanese. For those of you whose first language is not Japanese, the wording and expressions may seem more challenging than everyday Japanese. Additionally, many "trick questions" regarding Japanese traffic rules and etiquette are included, so careful preparation is necessary.

This book aims to help you navigate these trick questions and become familiar with the Japanese exam questions. We hope it will assist you in safely driving in Japan and living smoothly. Please use this book to successfully obtain your moped license on your first attempt.

Intellectual Property Business Laboratory
Director: HIDEHIKO YAMAMOTO

目　次

はじめに ･･･ 2

受験ガイド ･･･ 10

学科攻略ポイント ･････････････････････････････････････ 16

1．覚えておこう！標識と表示（18）

標識（20）

（1）規制標識 ･･･････････････････････････････････････ 20

（2）指示標識 ･･･････････････････････････････････････ 22

（3）警戒標識 ･･･････････････････････････････････････ 22

（4）案内標識 ･･･････････････････････････････････････ 24

（5）補助標識 ･･･････････････････････････････････････ 24

標示（26）

（1）規制標示 ･･･････････････････････････････････････ 26

（2）指示標示 ･･･････････････････････････････････････ 28

（3）路側帯 ･･･ 30

2．押えておきたい！交通用語の意味（32）

（1）道路に関する用語 ･･･････････････････････････････ 32

（2）車に関する用語 ･･･････････････････････････････････ 34

（3）道路の設備の用語 ･･･････････････････････････････ 38

（4）その他の用語 ･･･････････････････････････････････ 40

3．押さえておきたい！交通ルール（42）

（1）歩行者のそばを通行するとき ･･･････････････････････ 42

（2）車が通行するところ ･････････････････････････････ 44

（3）車が通行できないところ ･････････････････････････ 48

（4）乗車と積載の制限 ･･･････････････････････････････ 50

（5）信号機の種類 ･･･････････････････････････････････ 54

（6）緊急自動車の優先 ･･･････････････････････････････ 58

（7）交差点を通行する際の注意点 ･････････････････････ 60

（8）路線バスなどの優先 ･･･････････････････････････ 64

（9）信号のない交差点の優先順位 ･････････････････････ 66

4 目　次

Table of Contents

Introduction ·· 3

Exam Guide ·· 11

Strategies for Written Test ··· 17

1. Let's Remember! Signs and Markings (19)

Signs (21)

（1） Regulatory Signs ·· 21

（2） Directional Signs ·· 23

（3） Warning Signs ··· 23

（4） Guide Signs ·· 25

（5） Supplementary Signs ·· 25

Markings (27)

（1） Regulatory Markings ·· 27

（2） Directional Markings ··· 29

（3） Side Strips ··· 31

2. Let's Remember! Key Traffic Terminology (33)

（1） Terms Related to Roads ·· 33

（2） Terms Related to Cars ··· 35

（3） Terms Related to Road Facilities ·· 39

（4） Other Terms ··· 41

3. Let's Remember! Important Traffic Rules (43)

（1） When Passing by Pedestrians. ··· 43

（2） The Zone Where Vehicles Pass. ··· 45

（3） The Places Where Cars Are Not Permitted to Pass. ····················· 49

（4） Restrictions on Riding and Loading ··· 51

（5） Types of Traffic Signals ·· 55

（6） Priority for Emergency Vehicles ··· 59

（7） Precautions When Passing through an Intersection ······················ 61

（8） Priority for Bus Routes, etc. ··· 65

（9） Priority at Intersections without Traffic Signals ························· 67

(10) 踏切の安全な渡り方 ……………………………………68
(11) 追越し，追抜き ……………………………………70
(12) 追越しが禁止されている場合 ……………………74
(13) 追越し禁止場所 ……………………………………76
(14) 駐車禁止場所 ………………………………………80
(15) 駐停車禁止場所 ……………………………………82
(16) 駐停車の仕方 ………………………………………84
(17) おさらい！徐行する場所，いなければいけない時 …88
(18) 警察官などの手信号の意味 ………………………92

4．危険な場所での運転（94）

（1）緊急事態が起きたとき ……………………………94
（2）交通事故が起きたとき ……………………………96
（3）大地震が発生したとき ……………………………98

5．まちがえやすい数字のまとめ（100）

（1）積載制限 ……………………………………………100
（2）規制速度と法定速度 ………………………………102
（3）けん引するときの最高速度 ………………………104
（4）追越し禁止 …………………………………………104
（5）徐行 …………………………………………………106
（6）歩行者などの保護 …………………………………106
（7）合図を出すとき，場所 ……………………………106
（8）駐車禁止場所 ………………………………………106
（9）駐車禁止の場所と時間 ……………………………106
(10) 路側帯のある道路での駐停車 ……………………108
(11) 衝撃力・遠心力・制動距離 ………………………108
(12) 車の停止距離 ………………………………………110
(13) 制動距離 ……………………………………………110

6．イラスト問題の攻略法（112）

（1）イラスト問題を解く攻略法！ ……………………112
（2）イラスト問題ではここをチェックしよう！ ……114

（10）Safe Way to Cross a Railroad Crossing································69

（11）Overtaking And Passing ················71

（12）When Overtaking Is Prohibited ················75

（13）Places Where Overtaking Is Prohibited················77

（14）Places Where Parking Is Prohibited ················81

（15）Places Where Parking and Stopping Are Prohibited ················83

（16）Rules for Stopping or Parking ················85

（17）Review The Places and Situations Where You Must Driving at Reduced Speed! ·····89

（18）The Meaning of Hand Signals from Police Officers, etc. ················93

4．Driving in Hazardous Situations and Locations （95）

（1）In Case of Emergencies················95

（2）When a Traffic Accident Occurs ················97

（3）In The Case A Great Earthquake Occurs ················99

5．Summary of Confusing Numbers （101）

（1）Load Limit················101

（2）Regulatory Speed Limit and Statutory Speed Limit ················103

（3）Maximum Speed Limit When Towing ················105

（4）No Overtaking ················105

（5）Driving at Reduced Speed················107

（6）Pedestrians Protection, etc. ················107

（7）When to Give Turn Signals ················107

（8）No Parking Zone················107

（9）No Parking Zone and Time················107

（10）Parking and Stopping on Roads with Side Strips················109

（11）Impact Force, Centrifugal Force, Braking Distance ················109

（12）Stopping Distance of a Car ················111

（13）Braking Distance ················111

6．Strategies for Solving Illustration-based Questions （113）

（1）Strategies for Solving Illustration-based Questions! ················113

（2）Check Here in Illustration-based Questions! ················115

7．危険予測の練習問題（116）

模擬試験

第1回　模擬試験・解答解説……………………………………………122

第2回　模擬試験・解答解説……………………………………………138

第3回　模擬試験・解答解説……………………………………………154

第4回　模擬試験・解答解説……………………………………………170

第5回　模擬試験・解答解説……………………………………………186

第6回　模擬試験・解答解説……………………………………………202

第7回　模擬試験・解答解説……………………………………………218

第8回　模擬試験・解答解説……………………………………………234

あとがき…………………………………………………………………250

7. Practice Problems for Hazard Prediction (117)

Practice Tests

First Practice Test: Answers and Explanations ·· 123

Second Practice Test: Answers and Explanations ··· 139

Third Practice Test: Answers and Explanations ·· 155

Fourth Practice Test: Answers and Explanations ·· 171

Fifth Practice Test: Answers and Explanations ·· 187

Sixth Practice Test: Answers and Explanations ·· 203

Seventh Practice Test: Answers and Explanations ··· 219

Eighth Practice Test: Answers and Explanations ··· 235

Afterword ··· 251

受験ガイド

① 受験する場所

原付免許を受験する場合は，住所地を管轄する運転免許試験所（一部の地域では警察）に行き受験申請を行い，その後試験を受けられます。

② 受験資格

年齢が16歳以上である。

下記の人は原付免許の受験ができません。

1. 免許を拒否された日から起算して，指定された期間を経過してない人
2. 免許を保留されている人
3. 免許を取り消された日から起算して，指定された期間を経過してない人
4. 免許の効力が，停止または仮停止されている人
5. 政令で定める次の病気にかかってる人
   - 幻覚症状を伴う精神病者
   - 発作による意識障害や運動障害のある人
   - 自動車などの安全な運転に支障をおよぼすおそれのある人

③ 受験に持参するもの

1. 住民票または免許証	初めて免許を受ける人は住民票（本籍が記載されている）が必要。運転免許書を取得されている人は，その免許証が必要。
2. 証明写真	縦30ミリ×横24ミリ，無帽，無背景，胸上正面で6カ月以内に撮影したもの。裏に氏名，撮影年月日を記入。カラー，白黒どちらも可。
3. 運転免許申請書	運転免許試験場に用意されています。
4. 本人確認書類，印鑑	初めて受験する人は保険証，パスポート，学生証などが必要。印鑑は必要ない受験地もある。
5. 卒業証明書（ある人のみ）	指定自動車教習所の卒業者は卒業日から1年以内は技能試験が免除される。
6. 受験料	受験手数料，免許証交付手数料が掛かります。詳しい受験料は窓口で確認します。

10 受験ガイド

Exam Guide

① The location for taking the exam

If you are taking the motorized bicycle license exam, you can apply for the exam at the Driver's License Testing Center (or in some areas, the police station) that has jurisdiction over your residential address. After applying, you can take the exam.

② Eligibility for Taking the Exam

Must be aged 16 or over.

The following individuals are not eligible to take the motorized bicycle license exam:

1. Individuals who have not completed the specified period since being refused a license.
2. Individuals whose license is under suspension.
3. Individuals who have not completed the specified period since their license was revoked.
4. Individuals whose license's effectiveness is suspended or provisionally suspended.
5. Individuals suffering from the following illnesses specified by government ordinance:
   - Individuals with mental illness accompanied by hallucinations.
   - Individuals with consciousness or motor disorders due to seizures.
   - Individuals who may interfere with the safe driving of vehicles such as automobiles.

③ Items to Bring for the Exam

1. Resident Certificate or Driver's License	Those applying for a license for the first time need to bring their resident certificate (with their permanent address listed). For those who already have a driver's license, they need to bring that license.
2. Identification Photograph	Vertical 30mm x Horizontal 24mm, no hat, no background, taken within the last 6 months, chest up front view. Write your name and the date of the photo on the back. Color or black and white photos are both acceptable.
3. Driver's License Application Form	They are available at the Driver's License Testing Center.
4. Proof of Identity and Personal Seal (Hanko/ Inkan)	For first-time to take the test, documents such as insurance card, passport, or student ID may be required. Some test locations may not require a personal seal (Hanko/ Inkan).
5. Diploma (only for certain individuals)	Graduates of designated driving schools are exempt from the practical test for within one year from the date of graduation.
6. Examination Fee	There are examination fees and license issuance fees. Please check the exact examination fees at the counter.

Examination Guide 11

④　学科試験

1．出題範囲	原動機付自転車を運転するのに必要な交通ルール，安全運転の知識，原動機付自転車の構造や取扱など。
2．解答方法	問題を読んでマークシート方式の別紙の解答用紙に記入する。
3．制限時間	30分
4．出題内容	文章問題46問（各1点），イラスト問題2問（各2点）
5．合格基準	50点満点中45点以上で合格

⑤　適性試験の内容

1．視力検査
両眼で0.5以上あれば合格。片目が見えない人でも，もう片方の眼の視野が左右で150度以上，視力が0.5以上であれば合格。めがね，コンタクトレンズの使用も可。

2．色彩識別能力検査
赤，青，黄の色を見分ければ合格。

3．運動能力検査
車の運転に支障がなければ合格。義手や義足の使用も可。
＊身体に障害がある人は窓口で相談して下さい。

原付免許の試験には，実技試験がありません。その代わりに，原付講習を3時間受講することが，義務づけられています。

〈備考〉
・排気量50cc以下のバイクを「原付」，125cc以下のバイクを「第二種原動機付自転車（原付二種）」と呼びます。
・排気量125ccのバイクの運転には小型限定普通二輪免許が必要です。
・2025年11月より排ガス規制基準が強化され，原付の生産や販売が難しくなると予想されています。
・これに伴い，125cc以下のバイクを原付免許で運転できるようにする検討が行われています。
・ただし，最高出力が4キロワット以下で，原付と同等の加速力特性を持つバイクに限られます。
・125cc以下で最高速度が原付バイクと同等レベルのバイクであれば，原付免許で運転可能になります。
・2025年11月までに道交法改正でこの内容が施行される予定です。
・その際，ナンバープレートは50ccと同じ白色となり，二段階右折も必要になりま

④   **Written Test**

1. Scope of Questions	The knowledge of traffic rules and safe driving practices required for operating motorized bicycles, as well as the structure and handling of motorized bicycles.
2. Answering Method	Read the questions and mark your answers on the separate answer sheet provided in the mark sheet format.
3. Time Limit	30minutes
4. Contents of Questions	Text-based questions: 46 questions (1 point each), Illustration-based questions: 2questions (2 points each)
5. Passing Criteria	Passing score is 45 out of 50 points.

⑤   **The contents of the aptitude test**

1． Eye sight Test

Pass if the vision in both eyes is 0. 5 or higher. Even if one eye is blind, pass if the other eye has a visual field of 150 degrees or more in both horizontal directions and a vision of 0. 5 or higher. Glasses and contact lenses are also acceptable.

2． Color Perception Ability Test

Pass if you can distinguish between the colors red, blue, and yellow.

3． Physical Ability Test

Pass if there are no impediments to driving. The use of prosthetic limbs is also acceptable.
＊ Please consult at the counter if you have physical disabilities.

> **For the motorized bicycle license exam, there is no practical test. Instead, it is mandatory to attend a 3- hour motorized bicycle training course.**

Notes:

Motorcycles with an engine displacement of 50cc or less are called "Gentsuki" (mopeds), and those with an engine displacement of 125cc or less are referred to as "Class 2 Motorized Bicycles" (Gentsuki Ni-shu).

A "limited small-size motorcycle license" is required to operate a motorcycle with an engine displacement of 125cc.

Starting in November 2025, stricter emission regulations are expected to make the production and sale of mopeds more difficult.

In response, discussions are underway to allow motorcycles with an engine displacement of 125cc or less to be operated with a moped license.

Examination Guide   13

す。

・最高速度が原付よりも速い通常の 125cc バイクは，これまで通り小型限定普通二輪
免許が必要です。

However, this is limited to motorcycles with a maximum output of 4 kilowatts or less and with acceleration characteristics equivalent to that of mopeds.

If the motorcycle has an engine displacement of 125cc or less and a top speed equivalent to that of a moped, it can be operated with a moped license.

These changes are expected to be enacted through amendments to the Road Traffic Act by November 2025.

At that time, the license plate will be the same white color as that for 50cc motorcycles, and two-stage right turns will be required.

Regular 125cc motorcycles with a top speed higher than that of a moped will still require a "limited small-size motorcycle license" as before.

学科試験攻略ポイント

①問題は慌てず最後までしっかり読む！

文章問題には，まぎらわしい表現が出てきます。「〜である」「〜でない」などは，その意図を間違って解釈すると全く逆の解答になるので，文章はしっかり読みましょう。

②まぎらわしい法令用語の意味の違いに要注意！

「駐車」「停車」「追抜き」「追越し」などの法令用語は似ているので，要注意。このような言葉が出てきたら，その違いを意識して理解しましょう。

③時間があれば見直しを！

学科試験では，1問を解く時間がおよそ30秒前後です。自分では完璧と思っていても，思わぬ誤りに気付かないこともあります。時間に余裕があれば，見直しをして全体をチェックしましょう。

④イラスト問題はここに要注意！

イラスト問題には1つの問題につき3つの設問があり，1つでも間違えると得点になりません。配点はほかの問題の2倍なので，車や周囲の動きに気を配り，イラストをじっくり見て解答しましょう。

⑤分かる問題からどんどん解こう！

時間が限られているので，分からない問題で悩んでいると時間切れになってしまいます。分かる問題からどんどん解答して，分からない問題は空欄にせず，どちらかをマークしましょう。半分の確率で正解となるので，必ず空欄では終わらせてはいけません。

16　受験ガイド

Strategies for Written Test

① Don't panic, read the questions carefully until the end!

Text-based questions may contain tricky expressions. Phrases like 'is', 'is not', etc., can lead to completely opposite answers if misinterpreted, so make sure to read the text thoroughly.

② Pay close attention to the subtle differences in legal terminology!

Legal terms like 'parking', 'stopping', 'passing', 'overtaking', etc., may sound similar, so be careful. When you encounter such words, be conscious of their differences and understand them accordingly.

③ Review if you have time!

In the written test, you have approximately 30 seconds to answer each question. Even if you think you've answered perfectly, you may overlook unexpected errors. If you have spare time, review your answers and check the entire test.

④ Pay special attention to illustration-based questions!

Each illustration-based question has three sub-questions, and if you answer any of them incorrectly, you won't score any points. These questions carry double the points of other questions, so pay close attention to the movement of the car and its surroundings, and think carefully about the illustration before answering.

⑤ Start with the questions you can understand!

Since time is limited, spending too much time on questions you don't understand can lead to running out of time. Start by answering the questions you understand quickly, and if you encounter questions you don't understand, don't leave them blank—make an educated guess by marking one of the options. With a 50% chance of getting it right, you should never leave questions unanswered.

1. 覚えておこう！標識と標示

◎標識と標示の種類を覚えておこう！

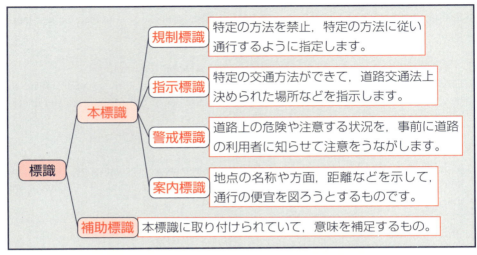

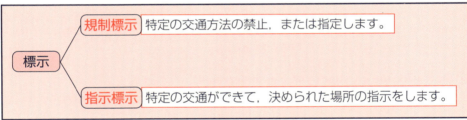

1. Let's remember! Signs and Markings

◎**Let's remember the types of signs and markings!**

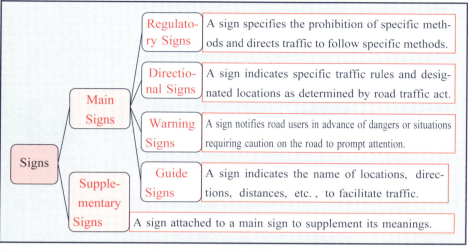

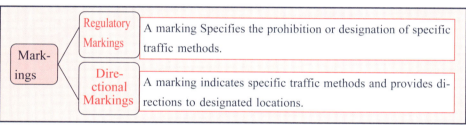

◎Let's Remember! Signs and Markings   19

標識（ひょうしき）

（1）規制標識（きせいひょうしき）

車両通行止め 自動車，原付自転車，軽車両は通行禁止。	**歩行者専用** 車は原則として通行できない。歩行者優先道路である。	**二輪の自動車以外の自動車通行止め** 二輪の自動車以外の自動車は通行禁止	**車両横断禁止** 車の横断禁止。	**指定方向外進行禁止** 矢印方向以外への車の進行禁止。

駐停車禁止 車は，8時から20時まで駐停車禁止。	**駐車禁止** 車は8時から20時まで駐車禁止。	**車両進入禁止** 車は，標識の示す方向からの進入禁止。	**転回禁止** 車は，転回（Uターン）してはならない。	**通行止め** 車，歩行者，路面電車は，通行禁止。

一時停止 車は，交差点の直前で一時停止しなければいけない。	**警笛区間** 車は，この標識のある区間内の指定場所で，警音器を鳴らすこと。	**一方通行** 車は，矢印の方向の反対方向に通行してはならない。	**追越し禁止** 車は，追越ししてはならない。	**追越しのための右側部分はみ出し通行禁止** 車は，道路の右部分にはみ出して追越ししてはならない。

Signs

(1) Regulatory Signs

No Entry for Vehicles	Pedestrian Paths	No Entry for Vehicles Other Than Two-wheeled Automobiles	No Crossing for Vehicles	No Entry for Vehicles in Directions Other Than the Arrow
No entry for automobiles, motorized bicycles, and light road vehicles	In principle, vehicles are not allowed to pass. It is a pedestrian priority road.	The passage of automobiles other than motorcycles is prohibited.	Vehicle crossings are prohibited.	It is prohibited for vehicles to travel in any direction other than the direction indicated by the arrow.

No Parking or Stopping	No Parking	No Entry for Vehicles	No turning back.	Road Closed
The sign below indicates no parking or stopping for vehicles from 8:00 to 20:00	The sign below indicates parking is prohibited for vehicles from 8:00 to 20:00	Vehicles are prohibited from entering in the direction indicated by the sign.	Vehicles must not make U-turns.	No entry for vehicles, pedestrians, and streetcars

Stop	Sounding Horn	One-way	No overtaking	No passing on The Right-hand Part of The Road for Overtaking
Vehicles must come to a stop just before the intersection.	Vehicles must sound horn within the section indicated by this sign.	Vehicles must not pass in the opposite direction of the arrow.	Overtaking is prohibited.	Vehicles must not stray to the right side of the road to pass.

（2）指示標識（しじひょうしき）

横断歩道	優先道路	安全地帯	駐車可	停止線
横断歩道であることを表している。	優先道路であることを表している。	安全地帯であることを表している。	駐車が可能であることを表している。	車両が停止する場合の位置を表している。

（3）警戒標識（けいかいひょうしき）

道路工事中	学校，幼稚園，保育所などあり	幅員減少	合流交通あり	踏切あり
前方の道路が工事中であることを表している。	周りに学校，幼稚園，保育所があることを表している。	道幅が狭くなっていることを表している。	この先に合流する道路があることを表している。	この先に踏切があることを表している。

ロータリーあり	十形道路交差点あり	車線数減少	二方向交通	落石のおそれあり
ロータリーがあることを表している。	十字道路交差点があることを表している。	車線数が少なくなることを表している。	対面通行の道路であることを表している。	落石の恐れがあることを表している。

(2) Directional signs

Pedestrian Crossing	Priority Road	Safety zone or Safety area	Parking Permitted	Stop Line
The sign indicates a pedestrian crossing.	The sign indicates a priority road.	The sign indicates a safety zone.	The sign indicates parking is permitted.	Stopping position

(3) Warning signs

Road Works	School, Kindergarten, Nursery, etc.	Road Narrows	Road Entry	Railroad Crossing
Road works ahead	The presence of a school, kindergarten, nursery, etc. in the vicinity.	The sign indicates that the road width is reduced.	Road entry ahead.	Railroad crossing ahead.
Roundabout	**Intersection**	**Lane reduction.**	**Two-way Traffic.**	**Falling Rocks.**
Roundabout ahead.	Intersection ahead.	Lane reduction ahead.	The sign indicates that the area is two-way traffic.	The sign indicates the risk of falling rocks.

(3) Warning signs   23

(4)案内標識（あんないひょうしき）

方面と距離（ほうめんときょり）
方面と距離を表している。

待避所（たいひじょ）
待避所であることを表している。

(5)補助標識（ほじょひょうしき）

始まり（はじまり）
本標識の規制区間がここから始まることを表している。

日・時間（ひ・じかん）
本標識が標示している規制の，適用される曜日や時間を表している。

(4) Guide Signs

Direction and distance	Shelter
The sign indicates direction and distance. 	The sign indicates the shelter ahead. 

(5) Supplementary signs

Beginning of the traffic regulation	Days and Times.
The sign indicates the beginning of the traffic regulation from here. 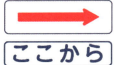 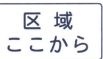	The sign indicates the days and the times when the traffic regulation applies. 

標示（ひょうじ）

（1）規制標示（きせいひょうじ）

駐車禁止	立ち入り禁止部分	転回禁止	最高速度
車は，駐車禁止。	車は，黄色の枠内へ入ることを禁止。	車は，回転（Uターン）禁止。	車両，路面電車の最高速度を表している。

路側帯	駐停車禁止	車両通行区分	停止禁止部分
歩行者と軽車両は通行可能で，幅0.75メートルを超える場合は路側帯に入って駐停車可能。	駐車と停車は禁止。破線は駐車禁止のみ。	車の種類によって通行位置が指定された車両通行帯を表している。	白色の枠内での車両，路面電車の停止が禁止。

優先本線車道	終わり
標示がある本線車道と合流する前の本線車道が優先道路であることを表している。	規制標示が示す交通規制の区間の終わりであることを表している。

Markings

(1) Regulatory markings

No Parking	No Entry Zone.	No U-turn	Maximum Speed Limit
Vehicles are prohibited from parking.	Vehicles are prohibited from entering the yellow framed area.	U-turns of vehicles are prohibited.	The maximum speed limit for vehicles and streetcars.

Side Strip	**No parking or stopping**	**Vehicular traffic lane**	**No Stopping Zone**
Pedestrians and light road vehicles are allowed to pass, and if the width exceeds 0.75meters, parking and stopping are permitted on the side strip.	Parking and stopping are prohibited. Dotted lines indicate no parking only.	Vehicle type-specific traffic lane	No stopping zone within the white frame for vehicles and streetcars.

Priority main lane for vehicles.	End of traffic regulations
The main lane before merging with the lane marked with a sign is the priority road.	The marking indicates the end of the traffic regulations designated by the regulatory markings.

(1) Regulatory markings

（2）指示標示（しじひょうじ）

路面電車停留場 路面電車の停留所であることを表している。 	**安全地帯** 安全地帯であることを表している。 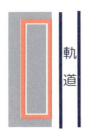	**右側通行** 車は，道路の右側部分にはみ出して通行できる。 	**二段停止線** 二輪車と二輪車以外の車の停止位置をそれぞれ表している。 
前方優先道路 前の道路が優先道路であることを表している。 	**自転車横断帯** 自転車が道路を横断できるところを表している。 	**横断歩道** 横断歩道であることを表している。 	**進行方向** 矢印の方向に進むことができる。 
停止線 車は，この位置で停止することを表している。 	**中央線** 道路の中央が中央線を示す。 		**安全地帯または路上障害物に接近** 安全地帯または路上障害物があって，接近していることを表している。

（2）Directive Markings

Streetcar stop	Safety Area	Keep Right	Two-Stage Stop Line
The marking indicates a streetcar stop	The marking indicates a safety area.	Vehicles may travel by straying on the right side of the road.	Two-stage stop line indicates the stopping position for two-wheeled vehicles and other vehicles respectively.

Priority Road Ahead	Bicycle Crossing	Pedestrian Crossing.	The Direction of Travel
Priority road ahead	The zone where bicycles can cross the road.	The marking indicates the crosswalk.	You can proceed in the direction of the arrow.

Stop Line	Center Line	Safety Zone or Obstacle on the Road Ahead.
The marking indicates stopping at this position.	Center line is a line that marks the center of the road and is typically drawn down the middle of the lanes.	The marking indicates approaching safety zone or obstacle on the road.

（2）Directive Markings　29

（3）路側帯（ろそくたい）

路側帯の表示は，規制標示であり，以下の3種類がある。

路側帯	駐停車禁止路側帯	歩行者用路側帯
歩行者と軽車両が通行できる。	路側帯に入って，車の駐車と停車が禁止。	路側帯に入って，車の駐停車や軽車両の通行禁止。

✲ ポイント！ ✲

　標識や標示はすべて，巻頭の「道路標識・標示一覧」に載せてあり，どれも出題される可能性があるので，しっかり覚えよう。自動車，原動機付自転車，歩行者のどれに対しての規制なのかを確認しよう。

【ここで例題】

(1) 本標識には規制，指示，警戒，案内，補助標識の5種類がある。
　解 ✕　補助標識を除いて，4種類ある。
(2) 標示には規制標示と案内標示の2種類がある。
　解 ✕　規制標示と指示標示の2種類である。
(3) 図1の路側帯のある道路では，車は路側帯の中に駐停車できる。
　解 ✕　図1の歩行者用路側帯では，車は路側帯の中に駐停車できない。
(4) 図2の標識のある交差点では，必ず一時停止しなければならない。
　解 ✕　図2は停止線の標識なので，車の停止位置を示すものであり，必ず一時停止する必要はない。

図1

図2

(3) Side Strips

The markings on the side strips are regulatory signs, and there are three types

Side Strips	The Side Strip Where Parking and Stopping Are Prohibited	Sidewalk for Pedestrians
Pedestrians and light road vehicles can pass.	No parking or stopping on the side strips.	No parking or stopping for vehicles and no passage for light road vehicles in the side strip.
路側帯 ｜ 車道	路側帯 ┆ 車道	路側帯 ‖ 車道

※ Point

All signs and markings are listed in the "Road Sign and Marking List" at the beginning, and each of them could be potentially tested, so make sure to memorize them thoroughly. Confirm which regulations apply to automobiles, motorized bicycles, and pedestrians.

【Exercise】

(1) There are 5 types of signs including regulation, instruction, caution, guidance and auxiliary signs.
   Incorrect × There are 4 types excluding auxiliary signs.
(2) There are 2 types of markings: regulation and guidance.
   Incorrect × There are 2 types of markings :regulation and instruction.
(3) On the road with the roadside in Figure 1, vehicles can park or stop in the side strip.
   Incorrect × On the sidewalk for pedestrian in Figure 1, cars cannot park or stop in the side strip.
(4) At the intersection with the sign in Figure 2, you must always come to a temporary stop.
   Incorrect × Figure 2 is a stop line sign indicating the position for vehicles to stop, and it is not necessary to come to a temporary stop.

Figure 1

Figure 2

2．押さえておきたい！交通用語の意味

◎基本の用語をしっかり理解し，覚えましょう。

（1）道路に関する用語（どうろにかんするようご）

路側帯（ろそくたい）
歩道がない道路で，道路標示によって区画された歩行者用の通路。

車両通行帯（しゃりょうつうこうたい）
「車線」や「レーン」ともいう。車が通行する部分。

路肩（ろかた）
道路の端から0.5メートルの帯状の部分。

歩道（ほどう）
歩行者の通行の為ガードレール，柵，縁石線などの工作物によって区分された部分。

2. Let's Remember Key Traffic Terminology!

◎ **Understand and Remember Basic Terminology!**

(1) Terms Related to Roads

Side Strip	Shoulder of A Road
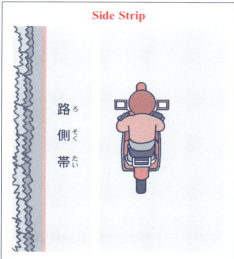	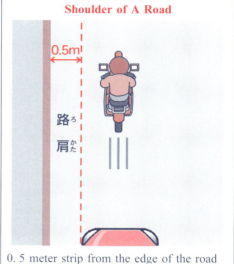
A pedestrian pathway demarcated by street markings on a street without sidewalks.	0.5 meter strip from the edge of the road
Vehicular Traffic Lane	**Sidewalk**
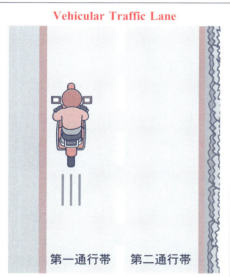	
It is also called lanes. The part of a road through which vehicles pass.	The area divided by guardrails, fences, curb lines, or other structures for pedestrian traffic.

(1) Terms Related to Roads   33

車道（しゃどう）	優先道路（ゆうせんどうろ）
	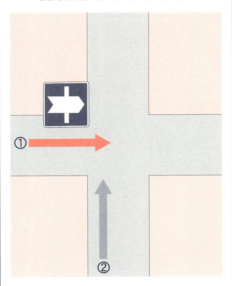
車両者用の通路と歩行者用の通路とが区別されている道路における車両用の通路。	「優先道路」の標識がある道路。交差点の中に中央線や車両通行帯がある道路。

（2）車に関する用語（くるまにかんするようご）

軽車両（けいしゃりょう）	歩行者（ほこうしゃ）	ミニカー（みにかー）

自転車，荷車，そり，また牛や馬のこと。原動機の付いていない車はおおむね軽車両。	道路を歩いている人。車いすや，小児用の車，二輪車のエンジンを止めて押している人も歩行者に含まれる。	排気量が 50cc 以下，または定格出力 0.6KW 以下の原動機を有する普通自動車のこと。

Roadway	Priority Road
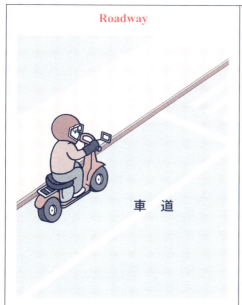	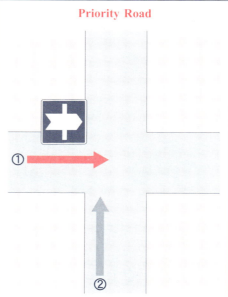
Vehicle pathways on streets where there is a distinction between pathways for vehicles and pedestrians.	Roads with Priority Road signs. A road with a center line or vehicular traffic lane in an intersection.

(2) Terms Related to Cars

Light Road Vehicle	Pedestrian	Minicar

Light road vehicles are bicycles or carts, sleds, oxen, and horses. Most non-motorized vehicles are included in the light road vehicles.	A person walking on the road. A person pushing a wheelchair, a child's car, or a two-wheeled vehicle with the engine turned off is also included as a pedestrian.	A standard motor vehicle with a displacement of 50 cc or less or a prime mover with a rated output of 0.6 KW or less.

車など（くるまなど）	車（くるま）
 車と路面電車。	 自動車，原動機付自転車，トロリーバス，軽車両。
路面電車（ろめんでんしゃ）	自動車（じどうしゃ）
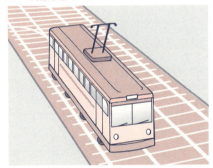 道路に敷かれた軌道に乗って走る電車。	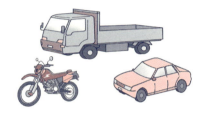 原動機を用いて，レールや架線によらないで運転する車。原動機付自転車や，自転車，車いす，歩行補助車は含まない。
原動機付自転車（げんどうきつきじてんしゃ）	緊急自動車（きんきゅうじどうしゃ）
 総排気量が50cc以下の二輪車。もしくは総排気量が20cc以下の三輪以上の車。	 赤色の警光灯をつけてサイレンが鳴っていたり，緊急のために運転中のパトカーや消防用自動車など。

Vehicles etc.  Vehicles and streetcars	**Vehicles**  Automobile, Motor-assisted Bicycle, Trolleybus, Light Road Vehicle
Streetcar 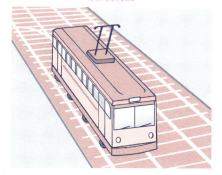 A streetcar running on tracks laid along a road.	**Automobile** 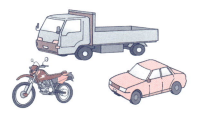 A vehicle powered by an engine and capable of operating without reliance on rails or overhead wires. This includes motorized bicycles, bicycles, and wheelchairs, and excludes walking aids.
Motorized bicycles  A motorized bicycle refers to a two-wheeled vehicle with a total displacement of 50 cc or less. It can also refer to a three-wheeled or more vehicle with a total displacement of 20 cc or less.	**Emergency Vehicle**  An emergency vehicle is a motor vehicle, such as a police car or fire-engine, equipped with flashing red lights and sirens that are used during emergencies to alert other road users and to allow the vehicle to proceed quickly through traffic.

(2) Terms Related to Cars

（3）道路の設備の用語（どうろのせつびのようご）

交差点（こうさてん） 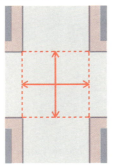 十字路やT字路などの2本以上の道路が交わる場所。	環状交差点（かんじょうこうさてん） 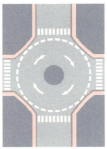 車両の通行部分が環状の交差点。右回りに車両が通行することが定められている。
立ち入り禁止部分（たちいりきんしぶぶん）  車が進入してはいけない表示部分。	標識（ひょうしき）  交通規制や道路の交通に関して指示を示す標示板。
信号機（しんごうき） 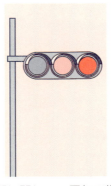 道路の交通に関して，電気で操作された灯火により，交通整理のための信号を標示するもの。	標示（ひょうじ）  道路の交通に関して，指示や規制などのためにペイントなどで路面に示された記号，線，文字。

(3) Terms Related to Road Facilities

Intersection	Roundabout
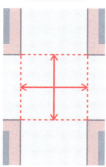	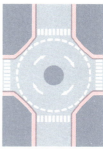
Intersection refers to a point where two or more roads intersect, such as a crossroads or a T-junction.	Roundabout is a type of intersection with a circular traffic flow pattern. Vehicles circulate around a central zone in a clockwise direction.
No Entry for Vehicles	**Signs**

Area where entry is prohibited for vehicles.	A sign that indicates traffic regulations or provides instructions regarding traffic on the road.
Traffic signal	**Markings**
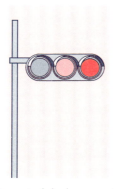	
Road traffic control device operated by electricity, displaying signals to regulate traffic flow.	Road markings, symbols, or letters painted on the road surface to provide instructions or regulations related to road traffic.

（4） その他の用語（そのたのようご）

総排気量（そうはいきりょう）

エンジンの大きさを示すのに用いられる数値。数値が大きければ，その車の馬力やトルクが大きくなる。

けん引（けんいん）

けん引自動車で故障車などをロープやクレーンで引っ張ったり，他の車を運んだりすること。

徐行（じょこう）

車が直ちに停止できそうな速度で走ること。

(4) Other terms

Total engine displacement

The value used to indicate the size of an engine. A larger number indicates greater horsepower and torque for the vehicle.

Towing

Towing involves using a towing vehicle to pull a disabled vehicle using a rope or crane, or transporting another vehicle.

Proceed at Reduced Speed

Driving at a speed at which the vehicle can immediately stop.

3．押さえておきたい！交通ルール

（1）歩行者のそばを通行するとき（ほこうしゃのそばをつうこうするとき）

歩行者などのそばを通行するとき

安全な間隔（1〜1.5ｍ以上）をあけるか，徐行しなければいけない。

停止中の車のそばを通行するとき

車のかげから人が飛び出したり，急にドアが開いたりするので十分に注意する。

停留所で停止中の路面電車のそばを通行するとき

乗降客や道路を横断する人がいなくなるまで後方で停止して待つ。

安全地帯のそばを通行するとき

歩行者がいるときは徐行して，いないときは徐行の必要はない。

3. Let's Remember Important Traffic Rules!

(1) When Passing by Pedestrians.

When passing by pedestrians etc.

You must keep **a safe distance (1-1.5 meters or more)** or drive **at a reduced speed.**

When passing by a stationary vehicle

Be sure to **exercise caution** as pedestrians may unexpectedly emerge from behind vehicles or car doors may suddenly open.

When passing by a stopped streetcars at a streetcar stop

Come to **a stop behind the streetcar** until passengers have boarded or alighted and pedestrians have crossed the road.

When passing near a safety zone

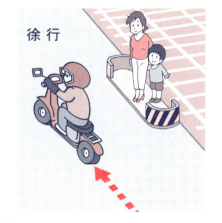

When pedestrians are present, drive **at reduced speed**; when they are not, there is **need to drive at reduced speed.**

以下のときは徐行して通行できる	子どもや身体の不自由な人のそばを通行するとき
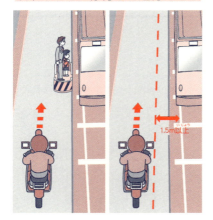	下のような人が通行しているときは，一時停止か徐行して，安全に通行できるようにする。 ①ひとり歩きしている子ども ②身体障害者用の車いすに乗っている ③盲導犬を連れている ④黄色か白色のつえをもっている ⑤通行に支障がある高齢者や身体障害者
安全地帯があるところ（乗降客がいてもいなくても徐行できる）	安全地帯がないところは，乗降者がいなければ路面電車との間に1.5ｍ以上の間をとって徐行できる。

（2）車が通行するところ（くるまがつうこうするところ）

車は，車道を通行する。
車は，道路の左側の部分を通行する。
車は，中央線があるときは，中央線から左側の部分を通行する。

標識，標示によって通行区分が指定された道路

自動車は，指定されている通行区分に従うが，原付は速度が遅いため，右折などやむを得ない場合以外は左側の通行帯を通行すること。

You can drive at reduced speed in the following situations.

When passing by children or individuals with **physical disabilities,**

when the following people are passing through

drive **at reduced speed** or come to **a stop** to ensure safe passage.

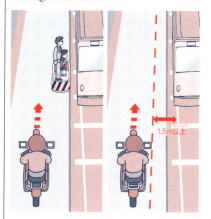

① A child walking alone
② Indibiduals riding in a wheelchair for people with physical disabilities
③ Individuals witha guide dog
④ Individuals carrying a yellow or white cane
⑤ Elderly persons or persons with physical disabilities whose movement may be impaired

You can drive at reduced speed where there is a safe zone (you can drive at reduced speed whether there are passengers **getting on or off or not**).	In areas without a safe zone, if there are no passengers boarding or alighting, you can drive **at reduced speed** while keeping **1.5 meters or more** from the streetcar.

(2) The Zone Where Vehicles Pass

Vehicles should pass **on the roadway.**
Vehicles should pass **on the left side** of the road.
Vehicles should pass **on the left side of the road when there is a center line.**

Roads where traffic lanes are designated by signs and markings.
Vehicles must adhere to the designated traffic lanes, while motorized bicycles, due to their slower speed, should generally use the **left** lane unless making a necessary right turn or similar maneuver.

---

(2) The Zone Where Vehicles Pass   45

右側にはみ出して通行できる場合（みぎがわにはみだしてつうこうできるばあい）

道路が一方通行になっている。

道路工事などで左側部分だけでは通行できないとき。

こう配の急な道路の曲がり角付近で、「右側通行」の標示があるとき。

左側部分の幅が6m未満の見通しのよい道路で、他の車を追い越そうとしているとき。

車両通行帯の通行（しゃりょうつうこうたいのつうこう）

＊車両通行帯のない道路では、車は左側を通行する。
＊2つの車両通行帯がある道路では、右側の通行帯は追い越しのためにあけておくので、左側を通行する。
＊3つ以上の車両通行帯がある道路では、原動機付自転車は速度が遅いため、右折などやむを得ない場合以外は最も左側の通行帯を通行する。

In cases where it is permissible, vehicles may pass while straying to the right side.

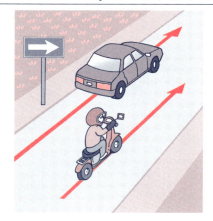

When the road is designated as *one-way.*

When road construction or similar activities prevent passage *on the left side.*

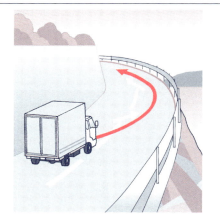

When there is a "*keep right*" sign near a sharp bend in the road with double lines.

When attempting to overtake another vehicle on a clear road with a width of *less than 6 meters* on the left side.

Passing in a lane designated for vehicles.

＊ On roads without designated lanes for vehicles, vehicles should pass *on the left*.

＊ On roads with two lanes for vehicles, the right lane is kept open for passing, so vehicles should pass *on the left*.

＊ On roads with three or more lanes for vehicles, motorized bicycles travel *in the leftmost* lane unless unavoidable, as they have slower speeds.

（3） 車が通行できないところ （くるまがつうこうできなところ）

標識や標示で禁止されている場所	歩行者専用通路
通行止め　　　車両通行止め  歩行者専用　　立ち入り禁止部分 	 沿道に車庫があるなどで**通行が認められた車**は通行できる。その場合は歩行者に注意して**徐行**する。
軌道敷内（きどうしきない）	歩道，路側帯，自転車道
 **右左折，横断，転回**をするために横切る場合や危険防止のため，やむを得ない場合や，**軌道敷内通行可**の標識によって通行が認められた自動車は，通行できる。	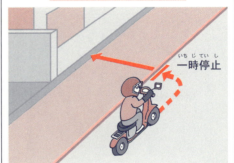 道路に面した場所に出入りのため，歩道や路側帯を**横切る**ことは可能。歩行者がいてもいなくてもその直前で**一時停止**すること。

渋滞などによる進入禁止 （じゅうたいなどによるしんにゅうきんし）

＊渋滞しているときなどは，交差点内や踏切や停止禁止部分の中で動きがとれなくなるおそれがある場合は進入禁止。

(3) The Places Where Cars Are Not Permitted to Pass.

The places where passage is prohibited by signs or markings	Pedestrian zone
 "Road Closed" "No Entry for Vehicles" "Pedestrian Paths" "No Entry Zone"	 **Vehicles that are permitted to pass**, such as those with garages along the road, are allowed to pass. In that case, be careful of pedestrians and **reduced speed**.
Inside the railway track area	Sidewalks, side strips, and bicycle lanes.
 Vehicles permitted to pass by the "**Railway Track Crossing**" signs, for **turning right or left, crossing**, or turning, or for safety measures, can pass.	It is possible to **cross** sidewalks or side strips to enter or exit places facing the road. Regardless of the presence of pedestrians, you must come to **a temporary stop** immediately before doing so.

Prohibition of entry due to congestion or similar situations.

∗ When there is congestion or similar situations, entry may be prohibited if there is a risk of being unable to move in **intersections, railroad crossings, or areas where stopping is prohibited**.

（4）乗車と積載の制限（じょうしゃとせきさいのせいげん）

原動機付自転車

乗車定員　運転者は1人のみ。

重量制限　重量制限は120kg以下。けん引できるのはリヤカー1台。その他のけん引は各都道府県によって異なる。

2人乗りは禁止

重量は120kg以下

積載物の制限

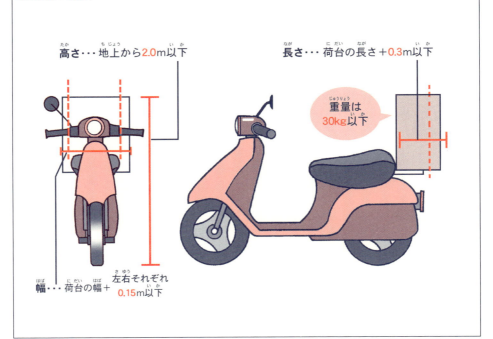

高さ・・・地上から2.0m以下

長さ・・・荷台の長さ＋0.3m以下

重量は30kg以下

幅・・・荷台の幅＋左右それぞれ0.15m以下

(4) Restrictions on Riding and Loading

Motorized bicycles

Passenger limit: Only **one** driver allowed

Riding with two people is prohibited.

The weight limit is **120 kg or less**. You can tow only one rear trailer. Other towing regulations may vary by prefecture.

The weight limit is 120 kg or less.

Cargo load limit

Height: **2.0** meters or less from the ground

Length: The length of the loading platform plus **0.3** meters

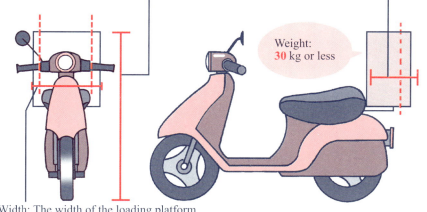

Weight: **30** kg or less

Width: The width of the loading platform plus **0.15** meters on each side

小型特殊自動車（とくしゅじどうしゃ）

乗車定員　運転者は **1人のみ**（運転者用以外に座席があるものは 2 人）
重量制限　500kg 以下

積載物の制限（せきさいぶつのせいげん）

長さ・・・**自動車の長さ×1.1 以下**
（長さ＋長さの 10 分の 1）

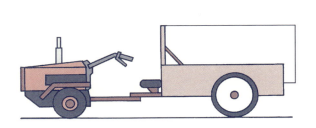
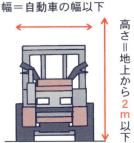

＊原付免許の学科試験は小型特殊自動車の試験と同じなので、小型特殊自動車についても覚えておこう。

Special Light Equipment

Passenger capacity: Only one person (vehicles with additional seats, other than for the driver, may have two persons)
Weight limit: 500 kg or less

Cargo limit

Length: **1.1 times the length of the vehicle or less**   Width: Equal to or less than the width of the vehicle

Height: **2 meters** or less from the ground

> * The written exam for a motorized bicycle license is the same as that for the special light equipment license, so it's a good idea to remember information about special light equipment as well.

(4) Restrictions on Riding and Loading

（5）信号機の種類（しんごうきのしゅるい）

青色の灯火	黄色の灯火	赤色の灯火
	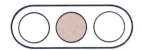	
車，路面電車は直進，右左折できる。（軽車両と二段階右折の原付は除く）	車，路面電車は停止位置から先へは進んではならない。（停止位置に近づいて安全に停止できない場合はそのまま進める）	車，路面電車は停止位置を越えて，進んではいけない。（すでに右左折している場合はそのまま進める）
青色矢印の灯火	黄色矢印の灯火
	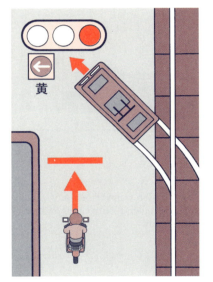
車は矢印の方向に進める。右の矢印の場合は転回もできる。（軽車両と二段階右折の原付は除く）	路面電車に対する信号。矢印の方向に進める。

(5) Types of Traffic Signals

Green Light	Yellow light	Red light
	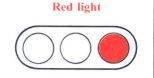	
Vehicles and streetcars may proceed **straight or turn right or left**. (Except for light road vehicles and motorized bicycles making a two-step right turn)	**Vehicles, streetcars** must not proceed beyond the stop line. (If it is not safe to stop near the stop line, they may proceed.)	**Vehicles and streetcars** must not proceed beyond the stop line; they must remain stopped. If they have already commenced turning, they may continue their turn.

Green arrow signal	Yellow arrow signal
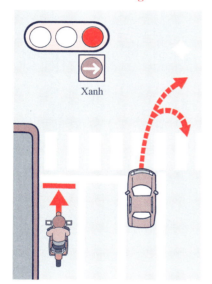	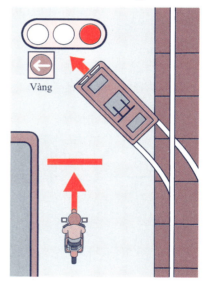
Vehicles can proceed in the direction of the arrow. In the case of a right arrow, vehicles can also **make a right turn**. (Excluding light road vehicles and motorized bicycles which have to make two-step right turns.)	Signal for **streetcars**. Proceed in **the direction of the arrow**.

(5) Types of Traffic Signals 55

赤色灯火の点滅

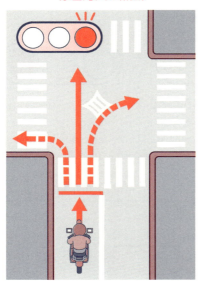

車，路面電車は停止位置で一時停止し，安全確認したあとに進める。

黄色灯火の点滅

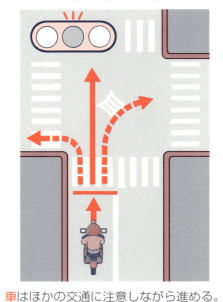

車はほかの交通に注意しながら進める。

原動機付自転車の二段階右折の標識があるところ

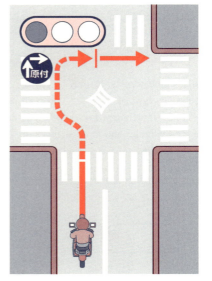

軽車両，原動機付自転車は，右折する位置まで進んで，その位置で向きを変更したあと，進むべき方向の信号機が青色になるまで待つ。

Flashing of red lights

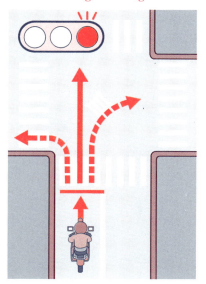

Vehicles and streetcars must come to a temporary stop at the stop position, proceed after confirming safety.

Flashing of yellow lights

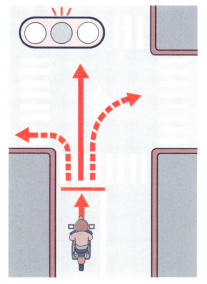

Vehicles should proceed with caution to other traffic.

The place where there is a sign for the two-stage right turn for motor-assisted bicycles

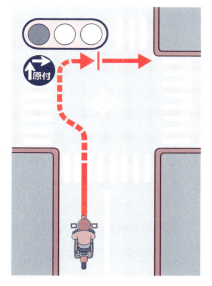

Light road vehicles and **motorized bicycles** should proceed to the position where they will make a right turn, change direction at that position, and wait until the traffic light for the direction they wish to proceed turns green.

(5) Types of Traffic Signals

（6）緊急自動車などの優先（きんきゅうじどうしゃなどのゆうせん）

緊急自動車とは
緊急用務の為に運転している消防用自動車やパトロールカー，救急用自動車などのこと。

交差点やその付近に緊急自動車が近づいてきた場合

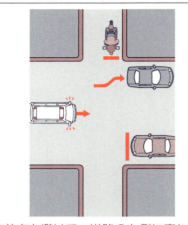

交差点を避けて，道路の左側に寄り，一時停止する。

一方通行の道路の場合，左側に寄ると妨げになるときは交差点を避けて，道路の右側に寄り一時停止する。

交差点やその付近以外で緊急自動車が近づいてきた場合

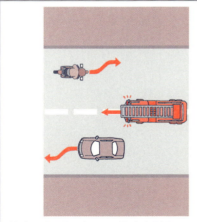

道路の左側に寄り，進路を譲る。

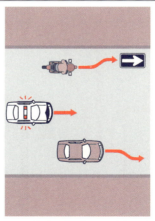

一方通行の道路の場合，左側に寄ると妨げになるときは，道路の右側に寄り，進路を譲る。

(6) Priority for Emergency Vehicles

What are the Emergency vehicles?
Vehicles such as fire engines, patrol cars, and ambulances that are being driven for emergency purposes.

In the case where an emergency vehicle approaches an intersection or its vicinity

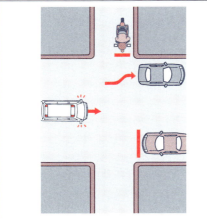

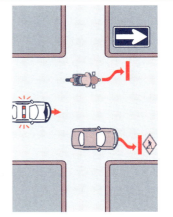

| Avoid the intersection, pull over to the left side of the road, and come to **a temporary stop**. | In the case of **a one-way street**, if moving to the left side would cause an obstruction, avoid the intersection, pull over to **the right side** of the road, and come to **a temporary stop**. |

In cases other than at or near intersections when an emergency vehicle approaches

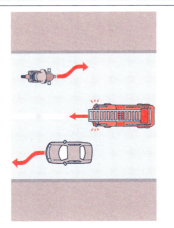

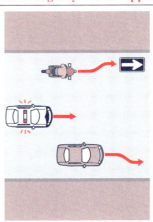

| Pull over to **the left side** of the road and yield the way. | In the case of **a one-way street**, if moving to the left side would cause an obstruction, move to **the right side** of the road and **yield the way**. |

（7）交差点を通行する際の注意点（こうさてんをつうこうするさいのちゅういてん）

右折の仕方 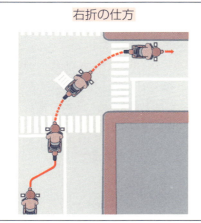	あらかじめ道路の中央に寄り，交差点のすぐ内側を徐行しながら通行する。
左折の仕方 	あらかじめ道路の左側に寄り，交差点の側端に沿って徐行する。
一方通行の場合 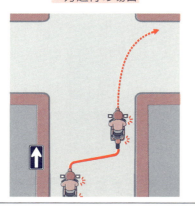	あらかじめ道路の右端に寄り，交差点の中心の内側を徐行しながら通行する。

60　3. 押さえておきたい！交通ルール

(7) Precautions When Passing through an Intersection

How to make a right turn	To make a right turn, move to **the center** of the road in advance, and then pass through the intersection at **a reduced speed**, staying just **inside** the intersection.
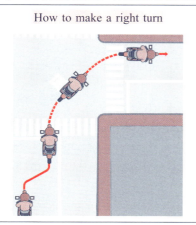	
How to make a left turn	Move to **the left side** of the road in advance. **Reduce spee**d and follow along **the edge** of the intersection.

In the case of a one-way street	Move to the right edge of the road in advance, and then pass through the center of the intersection at **a reduced speed**, staying **inside** the center.
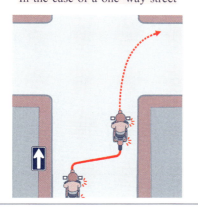	

環状交差点の通行の仕方（かんじょうこうさてんのつうこうのしかた）

環状交差点とは
下のように通行部分がドーナツ状の，右回りに通行する交差点

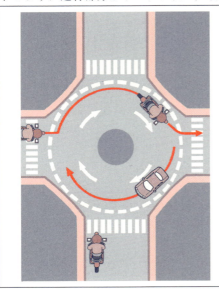

＊環状交差点に入るときは，徐行しながら環状交差点を通行している車両の通行を妨げない。
＊環状交差点内は，できる限り道路の左側端に沿って右回りに徐行する。
＊環状交差点から出るときは，出るところのひとつ前の出口通過直後に左折の合図を出し，交差点を出るまで合図を続ける。

二段階右折の仕方（にだんかいうせつのしかた）

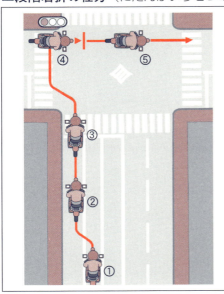

①あらかじめなるべく道路の左側に寄る。
②交差点の30ｍ手前で右折の合図を出す。
③青信号で徐行しながら交差点の向こう側まで直進する。
④この位置で停止し，右に向きを変えて，合図をやめる。
⑤前の信号が青になったら直進する。

How to pass through a roundabout

What is a roundabout?
A type of intersection where the traffic flows in a donut shape, moving in a clockwise direction.

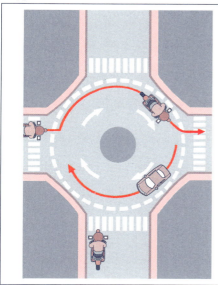

* When entering a roundabout, **reduce speed** and **do not hinder the passage** of vehicles already circulating in the roundabout.
* Inside the roundabout, proceed as close as possible to the left edge of the road, moving in **a clockwise direction**.
* When exiting a roundabout, **signal left** immediately after the exit just before the one you intend to take, and continue signaling **until you have exited the intersection**.

How to make a two-stage right turn

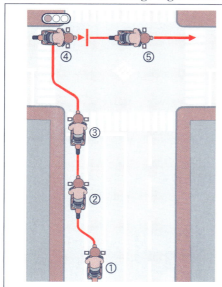

① Move as close as possible to **the left side** of the road in advance.
② Signal **a right turn 30 meters** before the intersection.
③ Proceed straight across the intersection at a reduced speed when **the signal is green**.
④ Stop at this position, turn **right**, and stop signaling.
⑤ Proceed **straight** when the signal ahead turns green.

(7) Precautions When Passing through an Intersection   63

二段階右折をしないといけない交差点（にだんかいうせつをしないといけないこうさてん）

①交通整理が行われており，3つ以上の車両通行帯がある道路の交差点。
②原動機付自転車の右折方法「二段階」の標識のある道路の交差点。

二段階右折してはいけない交差点（にだんかいうせつしてはいけないこうさてん）

①交通整理が行われており，車両通行帯が2つ以下の道路の交差点。
②交通整理が行われていない道路の交差点。
③原動機付自転車の右折方法「小回り」の標識がある道路の交差点。

（8）路線バスなどの優先（ろせんばすなどのゆうせん）

路線バスなどとは	路線バスの発進を妨げてはいけない
 路線バス，通学バス，通園バスのこと。	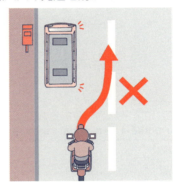 路線バスなどが発進の合図をしたとき，車は徐行，または一時停止をしてその発進を妨げてはいけない。ただし，急ハンドルや急ブレーキで避けなければいけないときは先に発進できる。

路線バス等優先通行帯では	路線バス等優先通行帯の標識と標示
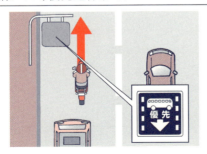	

64　3. 押さえておきたい！交通ルール

Intersections where a two-stage right turn is required

① Intersections of roads with traffic control and **three or more** lanes.
② Intersections on roads with a sign indicating **the two-stage** right turn method for motorized bicycles.

Intersections where a two-stage right turn is not allowed

① Intersections of roads with traffic control and **two or fewer** vehicle lanes.
② Intersections on roads without **traffic control**.
③ Intersections on roads with a sign indicating the "**tight turn**" right turn method for motorized bicycles.

(8) Priority of Route Buses etc.

What is the route buses?	Do not interfere with the departure of a bus.
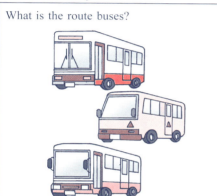 Route buses, school buses, and kindergarten buses	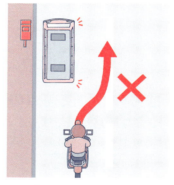 When route buses and similar vehicles signal to start, vehicles must **reduce speed** or come to **a temporary stop** and must not hinder their departure. However, vehicles can **proceed** first if they must avoid by taking sudden turns or applying emergency brakes.

In priority lanes for route buses and similar vehicles

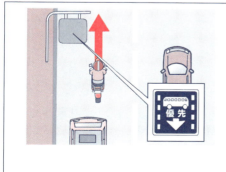	Signs and markings of priority lanes for route buses and similar vehicles 

(8) Priority of Route Buses etc. 65

＊路線バスなど以外に，自動車，原動機付自転車，軽車両も通行できる。
＊路線バスなどが近づいてきたときは，すぐに進路を譲る。
＊混雑時などで出られなくなるおそれがあるときは，初めから通行してはいけない。

バス専用通行帯では

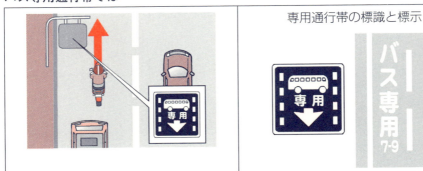

専用通行帯の標識と標示

＊原動機付自転車，小型特殊自動車，軽車両は通行できるが，それ以外の車は通行できない。
＊右左折や，工事などでやむを得ない場合は通行できる。

（9）信号のない交差点の優先順位（しんごうのないこうさてんのゆうせんじゅんい）

優先道路の標識がある道路。交差点の中まで中央線が引かれている道路。

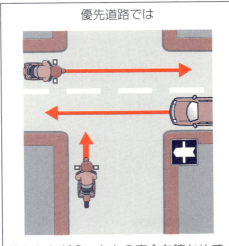

優先道路では

徐行しながら，左右の安全を確かめて，優先道路を通行する車の進行を妨げてはいけない。

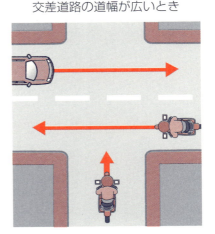

交差道路の道幅が広いとき

徐行しながら，左右の安全を確かめて，道幅が広い道路を通行する車の進行を妨げてはいけない。

* In addition to route buses and similar vehicles, **automobiles, motorized bicycles**, and light road vehicles can also use the lane.
* When route buses and similar vehicles approach, immediately **yield the way**.
* Do not **use the lane** from the start if there is a risk of being unable to exit during congestion.

In bus-only lanes

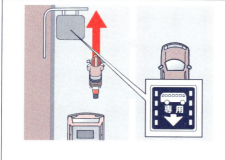

Signs and markings of exclusive lanes

* **Motorized bicycles, special light equipments, and light road vehicles** can pass, but other vehicles cannot.
* Turning right or left, or in unavoidable circumstances such as construction, passage is allowed.

(9) Priority at Intersections without Traffic Signals

Roads with priority road signs. Roads where the center line is drawn into the intersection.

On priority roads	When the width of the intersecting road is wide
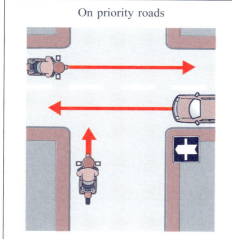	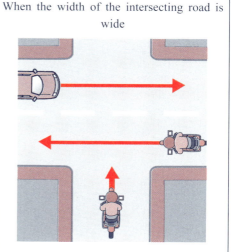
While driving at **reduced speed**, check the safety on both sides and **do not hinder the progress of** vehicles traveling on the priority road.	While driving at **reduced speed**, check the safety on both sides and **do not hinder the progress of vehicles** traveling on the wider road.

(9) Priority at Intersections without Traffic Signals 67

（10）踏切の安全な渡り方（ふみきりのあんぜんなわたりかた）

踏切の通過方法

踏切の直前で一時停止して，目と耳で安全確認して渡る。

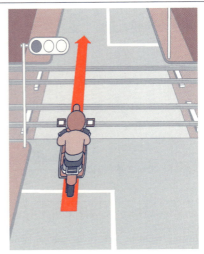

踏切に信号機があるところでは，その信号に従って渡る。青信号の場合は，安全確認するが，一時停止は不要。

遮断機がおりはじめているときは

遮断機が降りたり，警報機が鳴っているときは，踏切内に入ってはいけない。

前方の道路が渋滞しているときは

そのまま進入すると，踏切内で動きがとれなくなってしまうので，混雑時には踏切内に入ってはいけない。

(10) Safe Way to Cross A Railroad Crossing

Method of passing through a railroad crossing

 Stop temporarily just before the railroad crossing, and cross after confirming safety with both eyes and ears.	At railroad crossings with traffic signals, cross according to the signal. In the case of a green light, **confirm safety** but **a temporary stop** is not necessary.
When the railroad crossing barrier is starting to lower  **Do not enter** the railroad crossing when the barrier is down or the warning device is sounding.	When the road ahead is congested 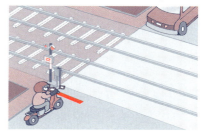 If proceeding would result in being unable to move within the railroad crossing, **do not enter** the crossing during congestion.

(11) 追越し，追抜き（おいこし，おいぬき）

追越し	追抜き
進路を変更して，進行中の前の車の前方に出ることをいう。	進路を変更せずに，進行中の前の車の前方に出ること。

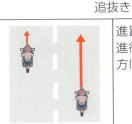

追越しの仕方（おいこしのしかた）

車の追越し	路面電車の追越し

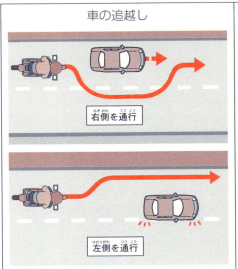
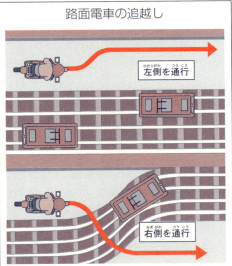

前の車の右側を通行するのが原則。ただし，前の車が右折のため中央に寄っているときは，その左側を通行する。

路面電車の左側を通行するのが原則。レールが左側に設けられている場合は除く。その場合は右側を通行する。

(11) Overtaking And Passing

Overtaking	Passing
The term refers to **changing** lanes to move ahead of the vehicle in front that is in motion.	Moving ahead of the vehicle in front that is in motion **without changing** lanes.

How to overtake

Overtaking a car	Overtaking a streetcar
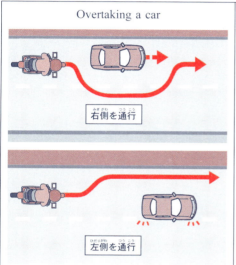	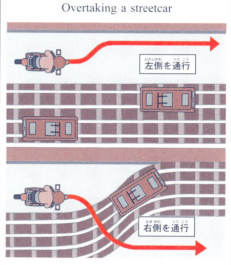
The principle is to pass on **the right side** of the car in front. However, if the car in front is moving to the center to turn right, then pass on **the left side**.	The principle is to pass on **the left side** of the streetcar, except when the rails are laid on the left side. In that case, pass on **the right side**.

(11) Overtaking And Passing 71

標識の意味（ひょうしきのいみ）

追越しのための右側部分は，はみ出し通行禁止  道路の右側部分にはみ出しての追越しは禁止されている（はみ出さなければ追越しできる）	追越し禁止  右側部分にはみ出さなくても，追越しは禁止されている。
追い越されるときの注意点 	①他の車に追い越されるときは，追越しが終わるまで速度を上げない。 ②追越しに十分な余地がない場合，できる限り左側に寄り，進路を譲る。

The meaning of traffic signs

No Passing on The Right-Hand Part of The Road for Overtaking Overtaking by straying onto the right side of the road is prohibited (overtaking is allowed if not straying).	No Overtaking  Even without straying to the right side, over- taking is prohibited.
Precautions when being overtaken 	① Do not increase speed until the overtaking by another vehicle is completed. ② If there is not enough room for overtak- ing, move as far to the left as possible and yield the way.

(11) Overtaking And Passing 73

（12）追越しが禁止されている場合（おいこしがきんしされているばあい）

 前の車が，右折などのため右側に進路変更しているとき。	 前の車が，自動車を追い越そうとしているとき。二重追越しという。前の車が原動機付自転車を追い越そうとしているときは，追越し可能。
 道路の右側に入って追越ししようとする場合に，反対方向からの車や路面電車の進行を妨げるとき。	 後ろの車が，自分の車を追い越そうとしているとき。

(12) When Overtaking Is Prohibited

When the vehicle in front is changing lanes to **the right side** for turning right or similar reasons.

When the vehicle in front is attempting to overtake another vehicle, it's called **double overtaking**. **Overtaking** is allowed when the vehicle in front is attempting to overtake a motorized bicycle.

When attempting to overtake by entering **the right side** of the road and **hindering the progress** of vehicles or streetcars coming from the opposite direction.

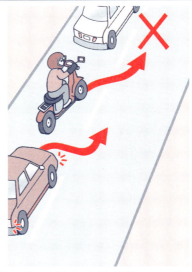

When the vehicle behind is attempting to overtake your vehicle.

（13）追越し禁止場所（おいこしきんしばしょ）

上り坂の頂上付近。

標識によって，追越しが禁止されている。

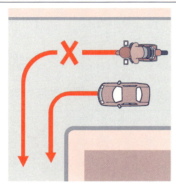
道路の曲がり角付近。

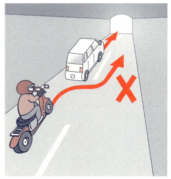
トンネルの中。（車両通行帯がある場合は可能）

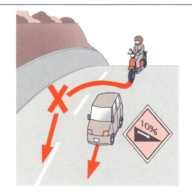
こう配の急な下り坂。（上り坂では禁止されていない。）

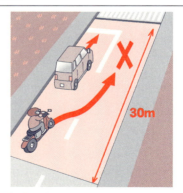
横断歩道や自転車横断帯とその手前から30m以内の場所。

(13) Places Where Overtaking Is Prohibited

Near the top of **an uphill slope**.

Overtaking is prohibited by **traffic signs**.

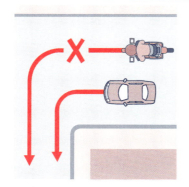

Near road **bends**.

Inside a tunnel. (Overtaking is possible if there are vehicle lanes.)

On **steep descents**. (It is not prohibited on uphill slopes.)

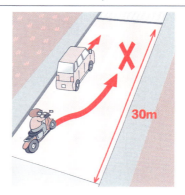

Within 30 meters before **a pedestrian crossing** or **bicycle crossing** and at the crossing itself.

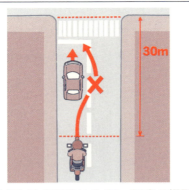 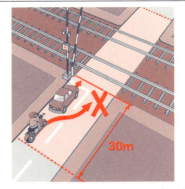

交差点とその手前から 30 m 以内の場所。
（優先道路を通行している場合は可能）

踏切とその手前から 30 m 以内の場所。

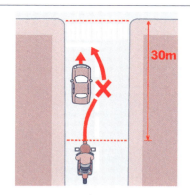	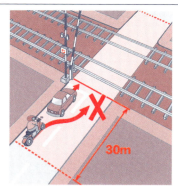
Within 30 meters before **an intersection** and at the intersection itself. (Overtaking is possible when traveling on a priority road.)	**Within 30 meters** before **a railroad crossing** and at the crossing itself.

(13) Places Where Overtaking Is Prohibited 79

(14) 駐車禁止場所（ちゅうしゃきんしばしょ）

駐車と停車
駐車とは車がすぐに運転できない状態の停止。
停車とはすぐに運転できる短時間の停止。

駐車禁止の標識や標示がある場所。

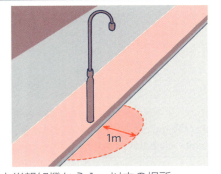

火災報知機から１m以内の場所。

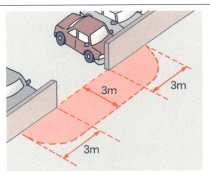

駐車場や車庫などの自動車用の出入口から３m以内の場所。

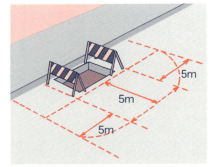

道路工事の区域の端から５m以内の場所。

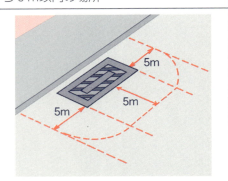

消火栓，指定消防水利の標識，消防用防火水槽の取り入れ口から５m以内の場所。

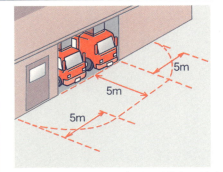

消防用機械器具の置き場や，消防用防火水槽などの出入口から５m以内の場所。

(14) Places Where Parking Is Prohibited

What is parking and stopping?
Parking refers to the stopping of a vehicle in a state where it cannot be immediately driven. Stopping refers to a short-term halt where the vehicle can be immediately driven.

 Places with no parking signs or markings.	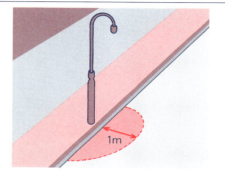 Within 1 meter of a fire alarm.
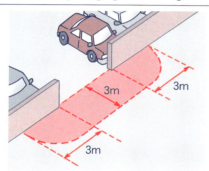 Within 3 meters of the entrance or exit of parking lots or garages for vehicles.	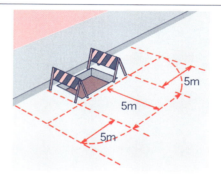 Within 5 meters of the edge of a road construction area.
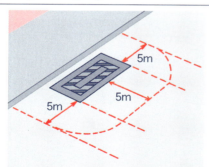 Within 5 meters of a fire hydrant, signs for designated firefighting water sources, or the intake of a fire protection water tank.	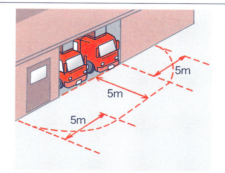 Within 5 meters of the storage area for firefighting equipment or the entrance or exit of a fire protection water tank.

(15) 駐停車禁止場所（ちゅうていしゃきんしばしょ）

駐車も停車も禁止されている場所

駐停車禁止の標識，標示がある場所。

坂の頂上付近や上がりも下りもこう配の急な坂。

トンネル内（車両通行帯のありなしにかかわらず禁止）。

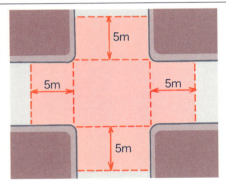

交差点とその端から5ｍ以内の場所。

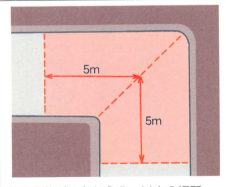

道路の曲がり角から5ｍ以内の場所。

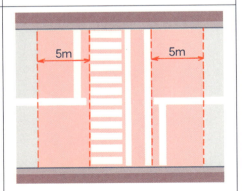

横断歩道や自転車横断帯とその端から5ｍ以内の場所。

(15) Places Where Parking and Stopping Are Prohibited

Places where both parking and stopping are prohibited

Places with signs or markings indicating no stopping and parking.

Near the top of a hill or on steep slopes, both ascending and descending.

Inside tunnels (prohibited regardless of whether there are vehicle lanes or not).

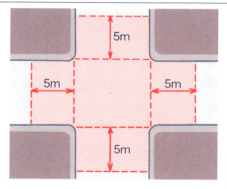

Within 5 meters of an intersection and its edges.

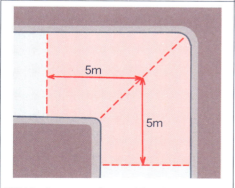

Within 5 meters of a road bend.

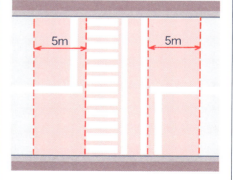

Within 5 meters of a pedestrian crossing or bicycle crossing and its edges.

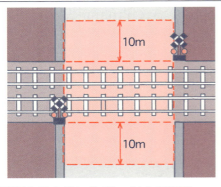
踏切とその端から 10 m 以内の場所。

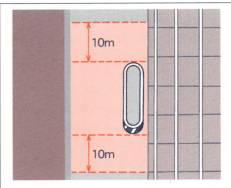
安全地帯の左側とその前後 10 m 以内の場所。

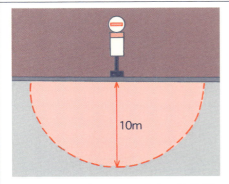
運行時間内のバス，路面電車の停留所の標示板から 10 m 以内の場所。

（16）駐停車の仕方（ちゅうていしゃのしかた）

歩道や路側帯のない道路の場合	路側帯のある道路の場合	
 道路の左端に沿う。	路側帯が幅 0.75 m 以下場合，車道の左端に沿う。	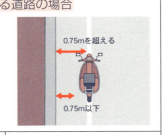 路側帯が幅 0.75 m を超える場合，路側帯に入り，左側に 0.75 m 以上の余地をあける。

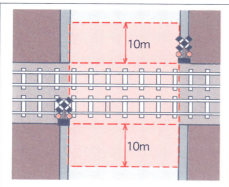

Within 10 meters of a railroad crossing and its edges.

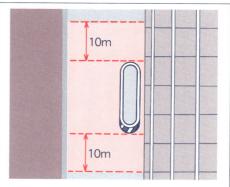

Within 10 meters before and after the left side of a safety zone.

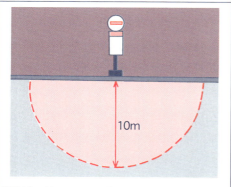

Within 10 meters of the bus and streetcar stops during their operating hours.

(16) Rules for Stopping or Parking

In the case of roads without sidewalks or side strips	In the case of roads with side strips	
		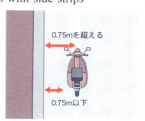
Park along the left side of **the road**.	If the shoulder is **less than 0.75 meters wide**, park along the left edge of the roadway.	If the shoulder is **wider than 0.75 meters**, enter the side strip and leave at least 0.75 meters of space on the left side.

(16) Rules for Stopping or Parking   85

歩道のある道路の場合	*2本線の路側帯がある道路の場合は，中に入ってはならない	
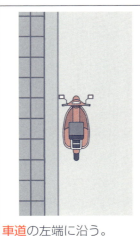	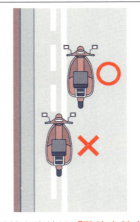	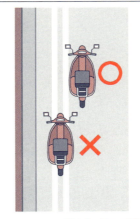
車道の左端に沿う。	破線と実線は「駐停車禁止路側帯」	実線2本は「歩行者用路側帯」

無余地駐車の禁止
・車の右側の道路上に3.5m以上の余地がない場所では，駐車禁止。
・標識によって余地が指定されている場所では，車の右側にその長さ以上の余地をあけなければならない。

余地がなくても以下の場合は駐車できる
・荷物の積み下ろしを行う場合で，運転者がすぐに運転できる場合。
・傷病者の救護のため，やむを得ない場合。

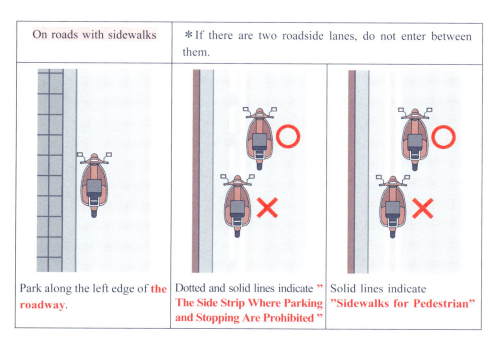

On roads with sidewalks	*If there are two roadside lanes, do not enter between them.	
Park along the left edge of **the roadway**.	Dotted and solid lines indicate " **The Side Strip Where Parking and Stopping Are Prohibited** "	Solid lines indicate "**Sidewalks for Pedestrian**"

No parking in areas without space
- In areas where there is no space of 3. 5 meters or more on the right side of the road, parking is prohibited.
- In areas where the space is designated by a sign, you must leave a space of at least that width on the right side of the vehicle.

You can park even if there is no space in the following cases:
- In the case **unloading luggage** and the driver can immediately drive.
- In unavoidable cases for **the rescue of the sick or injured**.

(16) Rules for Stopping or Parking 87

（17）おさらい！徐行する場所，しなければいけない時

◎徐行する場所，しなければいけない時をもう一度，確認しよう！

① 徐行の標識があるところ	② 左右の見通しがきかない交差点（*例外　信号機がある交差点，優先道路を通行している場合）

③ 上り坂の頂上付近，こう配の急な下り坂	④ 道路の曲がり角付近

⑤ 交差点で右左折する場合	⑥ ぬかるみ，水溜まりの場所を通行する場合
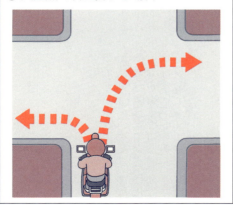	

(17) Review The Places and Situations Where You Must Driving at Reduced Speed!
◎ **Let's review the places and situations where you must reduce speed again!**

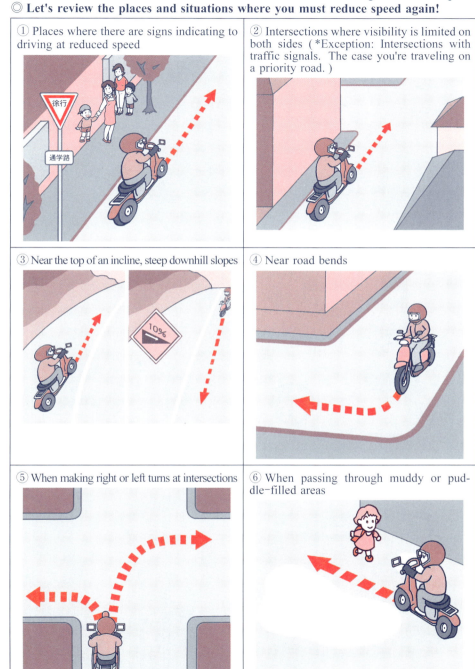

⑦ 優先道路または道幅の広い道路に進入する場合

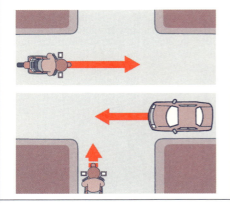

⑧ 道路外に出るため，右左折をする場合

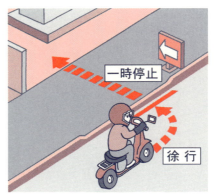

⑨ 歩行者用道路の許可を受けて通行する場合

⑩ 歩行者のいる安全地帯の横を通過する場合

⑪ 乗り降りのため，停車中の通学，通園バスの横を通行する場合

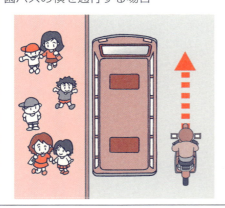

⑫ 乗り降りのいない停止中の路面電車との間隔が1.5m以上とれる場合

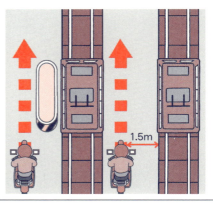

⑦ When entering a priority road or a wide road

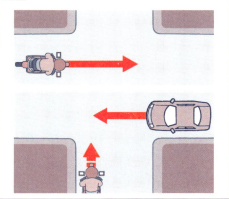

⑧ When making a right or left turn to exit the road

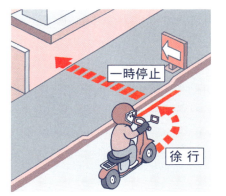

⑨ When traveling with permission on pedestrian pathways

⑩ When passing by the side of a safety zone where pedestrians are present

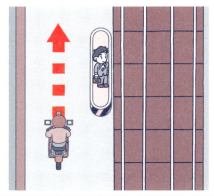

⑪ When passing by a stopped school or nursery school bus for boarding and alighting

⑫ When there is 1.5 meters or more between the stopped streetcar and where boarding or alighting is not occurring

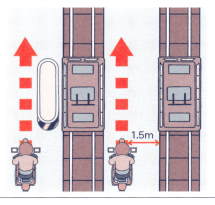

(17) Review The Places and Situations Where You Must Driving at Reduced Speed!

（18）警察官などの手信号の意味

＊**警察官などとは…警察官と交通巡視員のことをいう**
警察官などの手信号や灯火信号の意味

腕を横に水平に上げている場合	腕を垂直に上げている場合
 平行する交通…青色の灯火と同じ 対面する交通…赤色の灯火と同じ	 平行する交通…黄色の灯火と同じ 対面する交通…赤色の灯火と同じ
灯火を横に振っている場合	灯火を頭上に上げている場合
 平行する交通…青色の灯火と同じ 対面する交通…赤色の灯火と同じ	 平行する交通…黄色の灯火と同じ 対面する交通…赤色の灯火と同じ

信号機の信号と警察官などの手信号や灯火信号が，異なる場合は**警察官など**の信号に従う。

(18) The Meaning of Hand Signals from Traffic Police Officers

* **Traffic police officers refer to police officers and traffic wardens.**
The meaning of hand signals or light signals from traffic wardens

When the arm is raised horizontally	When the arm is raised vertically

If you are on a street parallel to police officer: same as blue light If you are on a street facing to police officer: same as red light	Parallel: same as yellow light Facing: same as red light
When waving the light horizontally	When raising the light above the head
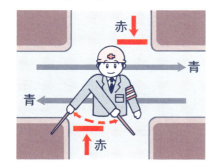	
Parallel: same as blue light Facing: same as red light	Parallel: same as yellow light Facing: same as red light

If the signals from traffic lights and those from traffic police officers or light signals differ, follow the signals from **the traffic police officer**.

4．危険な場合や場所での運転（きけんなばあいやばしょでのうんてん）

（1）緊急事態がおきたとき（きんきゅうじたいがおきたとき）

① 対向車と正面衝突のおそれがある場合

できる限り左側に寄り，警音器とブレーキを使う。道路外に危険がなければ，ためらうことなく道路外に出る。

② 後輪が横滑りした場合

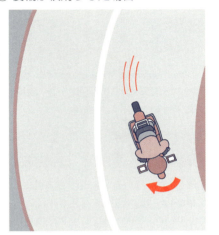

ブレーキをかけずに，スロットルをゆるめる。同時に後輪が滑る方向にハンドルを切り，車体を立て直す。

③ 下り坂でブレーキが効かなくなる場合

減速チェンジの後に，エンジンブレーキを効かせながら，減速します。それでも減速しなければ，道路外に停止を試みる。

④ スロットルが戻らない場合

ただちに点火スイッチを切る。エンジンの回転を止めて，ブレーキをかけながら道路の左端に停止する。

4. Driving in Dangerous Situations or Locations

(1) In Case of Emergencies

① When there is a risk of head-on collision with on-coming traffic

As much as possible, move to **the left side**. Use **the horn and brakes**. If there is no danger off the road, don't hesitate to leave the road.

② If the rear wheels skid sideways

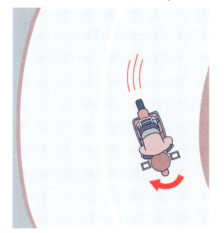

Without applying the brakes, release **the throttle**. Simultaneously, steer the handlebars in **the direction the rear wheels are sliding** to regain control of the vehicle.

③ When the brakes fail on a downhill slope

After **downshifting** to **reduce speed**, use engine braking while decelerating. If this doesn't work, attempt to stop **off the road**.

④ If the throttle doesn't return

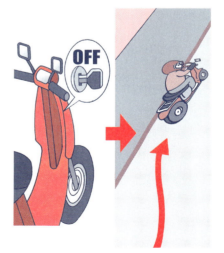

Immediately turn off **the ignition switch**. Stop the engine and apply the brakes while stopping at **the left edge** of the road.

⑤ 走行中にタイヤがパンクした場合	しっかりとハンドルを握り，車体をまっすぐに保つ。ブレーキを継続的にかけながら，道路の左端に停止する。

（2）交通事故がおきたとき（こうつうじこがおきたとき）

① 事故の続発を防止	② 負傷者の救護
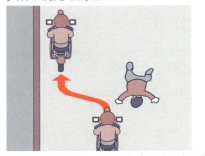	
事故の続発防止のため，安全な場所に車を移動する。	負傷者がいる場合は，すぐに救急車を呼んで応急救護措置を行う。
③ 警察官へ報告	

警察官に事故が発生した場所や状況など報告する。	

⑤ If a tire punctures while driving | Grip the steering wheel firmly and keep the vehicle straight. Apply the brakes **continuously** while stopping at **the left edge** of the road.

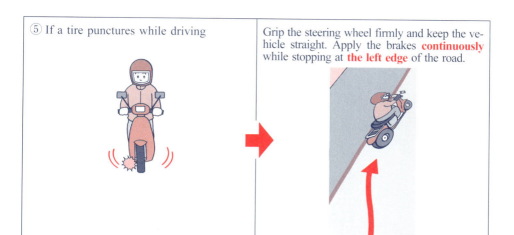

(2) When a Traffic Accident Occurs

① Preventing further accidents

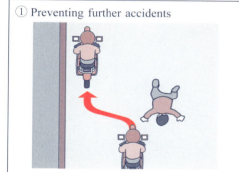

To prevent further accidents, move the vehicle to a safe location.

② Rendering aid to the injured

If there are injured individuals, immediately call for **an ambulance** and administer **first aid** measures

③ Report the incident to the police.

Report the location and circumstances of the accident to **the police officer**.

（３）大地震が発生した時（だいじしんがはっせいしたとき）

① 急ブレーキを避けて，車を停止する。
ラジオや携帯などで地震情報を確認。

② 車を置いて避難する場合は，できる
限り道路外の場所に車を移動する。

③ やむを得ず道路上に車を置いて避難
した場合は，鍵をつけたままにするか，
誰でも移動できるようにする。

④ 避難する場合は，混雑を防ぐため，
車での移動は避ける。

(3) In The Case A Great Earthquake Occurs

① Avoid **sudden braking** and bring the vehicle to **a stop**. Check earthquake information by radio, mobile phones, etc.

② If evacuating and leaving the car behind, try to **move** the vehicle to a location off the road, if possible.

③ If evacuating and leaving the vehicle parked on a necessary passage, **leave the keys** in the vehicle for accessibility by anyone who needs to **move** it.

④ When evacuating, avoid **traveling** by vehicles to prevent congestion.

5．まちがえやすい数字のまとめ

　色んな数字に関する問題が出題されるのでテーマにわけて，覚えよう！「○m以内」「時速○km」などの数字の問題や，「手前○m以内」と「前後○m以内」の違いに気をつけて，「以上」「以下」「以内」「越える」「未満」などの意味も覚えておこう。

（1）積載制限（せきさいせいげん）

● 普通自動車の積載制限は，地上からの高さ **3.8m以下**，自動車の長さ× **1.1m以下**，自動車の幅以下
● 原動機付自転車の積載制限は，地上からの高さ **2.0m以下**，積載装置の長さ+ **0.3m以下**，積載装置の幅+それぞれの左右 **0.15m以下**
● 原動機付自転車の最大積載量は **30kg**（リヤカーでのけん引き時は **120kg** まで。*リヤカーのけん引の許可は都道府県によって異なる）

区分	積載物の大きさと積載の方法
普通自動車 準中型自動車 中型自動車 大型特殊自動車 大型自動車	自動車の長さ×1.1m以下　　自動車の幅以下 以下3.8m 三輪の普通自動車，総排気量660cc以下の普通自動車は高さ2.5m以下
大型特殊二輪車 普通自動二輪車	積載装置の長さ+0.3m以下　　積載装置の幅+左右0.15m以下 以下2.0m
原動機付自転車	同上

100　5．まちがえやすい数字のまとめ

5. Summary of Confusing Numbers:

Let's organize and memorize various numbers-related questions by theme! Be mindful of differences like "within ◯ meters ahead" and "within ◯ meters before or after. " Also, remember the meanings of terms like "within, " "more than, " "less than, " and "up to, " commonly used in phrases like "within ◯ meters" or "at a speed of ◯ kilometers per hour. "

（1） Load Limit

● The load limit for regular motor vehicles is a height of **3. 8 meters or less** from the ground, **not exceeding 1. 1times** the length of the vehicle, and **not exceeding the width of the vehicle**.

● The load limit for motorized bicycles is a height of **2. 0 meters or less** from the ground, the length of the loading device plus **0. 3meters or less**, and the width of the loading device plus **0. 15 meters or less** on each side.

● The maximum load capacity for motorized bicycles is **30 kg** (up to 120 kg when towing with a trailer). *Permission to tow with a trailer may vary depending on the prefecture.

Category	Size and Loading Method of Cargo	
Regular Motor Vehicles **Sub-medium Vehicles** **Medium Vehicles** **Special Heavy Equipments** **Large Vehicles**	**Not exceeding 1. 1 times** the Length of the vehicle	Equal to or less than the width of the vehicle 3. 8 meters or less 3.8m Three-wheeled regular motor vehicles and regular motor vehicles with a total displacement of 660 cc or less have a height limit of 2.5 meters or less.
Large-size motorcycles **Regular motorcycles**	Length of the loading device **+ 0. 3 meters or less**	Width of the loading device **+ 0. 15 meters on either side or less** **2. 0 meters or less**

（1） Load Limit　101

| 小型特殊自動車 | 自動車の長さ×1.1m以下　自動車の幅以下　2.0m以下 |

（2） 規制速度と法定速度 （きせいそくどとほうていそくど）

● 標識や標示で最高速度が指定されている道路は，その**規制速度**内で運転する。
● 標識や標示で最高速度が指定されていない道路は，以下の**法定速度**内で運転する。

区分	最高速度
大型，中型，準中型乗用自動車 大型，中型，準中型貨物自動車 大型特殊自動車 けん引自動車 普通貨物自動車 普通乗用自動車 大型自動二輪車 普通自動二輪車 総排気量660cc以下の自動車 ミニカー	時速**60**km
原動機付自転車	時速**30**km

102　5. まちがえやすい数字のまとめ

Motorized bicycle	The same as above
Special light equipment	Not exceeding 1.1 times the length of the automobile — The width of the automobile or less

（2） Regulatory Speed Limit and Statutory Speed Limit

● Roads where the maximum speed limit is designated by signs or markings are driven within **that regulatory speed** limit.

● On roads where the maximum speed is not designated by signs or markings, driving is within the following **statutory speed** limits.

Category	Maximum Speed Limit
Passenger Cars: Large, Medium, Semi-medium Vehicles Cargo Trucks: Large, Medium, Semi-medium Vehicles Large Special Equipment Tow Trucks Regular Cargo Trucks Regular Motor Vehicles Large Motorcycles Regular Motorcycles Vehicles with total displacement of 660 cc or less Mini Cars	**60 km/ h**
Motorized bicycle	**30 km/ h**

（3）けん引するときの最高速度（けんいんするときのさいこうそくど）

車両総重量2000kg以下の故障車などを，その3倍以上の車両総重量の車でけん引するとき。		時速 40km
それ以外の場合で故障車などをけん引するとき。		時速 30km
原動機付自転車や125cc以下の普通自動二輪車でけん引するとき		時速 25km

（4）追越し禁止（おいこしきんし）

●何m以内なのかをしっかり覚えて，前後か手前かについても整理しておこう！

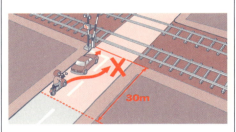

踏切とその手前から30m以内の場所は追越し禁止。

交差点とその手前から30m以内の場所は追越し禁止。（優先道路を通行しているときは例外）

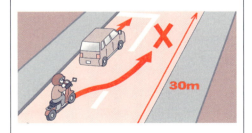

横断歩道，自転車横断帯とその手前から30m以内の場所は追越し禁止。

(3) Maximum Speed Limit When Towing

When towing a disabled vehicle with a gross vehicle weight of 2000 kg or less using a vehicle with a gross vehicle weight that is three times or more that of the disabled vehicle:		40 km/h
When towing a disabled vehicle in any other case:		30 km/h
When towing with a motorized bicycle or a regular motorcycle with an engine displacement of 125 cc or less:		25 km/h

(4) No Overtaking.

● Let's review the distances: Remember to understand whether it's "within" or "before/after" a certain distance.

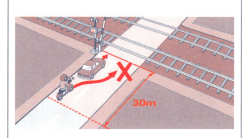

Passing is prohibited **within 30meters** before and at the railroad crossing.

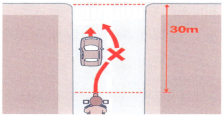

Passing is prohibited **within 30meters** before and at intersections, with an exception when driving on priority roads.

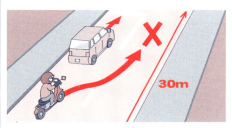

Passing is prohibited **within 30meters** before and at the pedestrian crossings, bicycle crossings, and their preceding areas.

（5） 徐行（じょこう）

●徐行とは

車がただちに停止できるような速度で進行すること	ブレーキ操作から1m以内で停止できる速度	おおむね時速10km以下の速度

（6） 歩行者などの保護（ほこうしゃなどのほご）

●歩行者や自転車のそばを通行するときは1〜1.5m以上の安全な間隔をあける。

（7） 合図を出すとき，場所（あいずをだすとき，ばしょ）

●右左折，転回の合図は30m手前で出す（環状交差点は除く）。
●進路を変えるときの合図は約3秒前に出す。
●環状交差点を出るときは直前の出口の側方を通過したとき（入るときは合図しない）。

（8） 駐車禁止場所（ちゅうしゃきんしばしょ）

●道路工事の区域の端から5m以内の場所
●火災報知機から1m以内の場所
●自動車専用の出入口（駐車場や車庫）から3m以内の場所
●消防用機械器具の置き場や消防用防火水槽，これらの道路に接する出入口から5m以内の場所
●消火栓，指定消防水利の標識がある位置や，消防用防火水槽の取入口から5m以内の場所

（9） 駐停車禁止の場所と時間（ちゅうていしゃきんしのばしょとじかん）

●荷物の積み下ろしのために5分を超えると，駐車。5分以内なら停車
●交差点とその端から5m以内の場所
●道路の曲がり角から5m以内の場所
●横断歩道，自転車横断帯とその端から前後5m以内の場所
●運行時間中のバス，路面電車の停留所の標識板（柱）から10m以内の場所
●踏切とその端から前後10m以内の場所
●安全地帯の左側とその前後10m以内の場所

(5) Driving at Reduced Speed

● Driving at Reduced Speed:

Driving at a speed at which the vehicle can come to an immediate stop.	A speed at which the vehicle can come to a stop **within 1 meter** after braking.	A speed generally **below 10 km/ h**.

(6) Protection of pedestrians, etc.

● When passing pedestrians or bicycles, keep a safe distance of **1 to 1. 5 meters** or more.

(7) The time and location to signal

● Right and left turns, and U-turn signals should be given **30 meters before** (except for roundabout).

● When changing lanes, signal about **3 seconds in advance**.

● When leaving a roundabout, signal when passing the side of the immediate exit (do not signal when entering).

(8) No parking zones

● **Within 5 meters** of the edge of a road construction zone

● **Within 1 meter** of a fire alarm device

● **Within 3 meters** of a dedicated entrance or exit for automobiles (such as parking lots or garages)

● **Within 5 meters** of fire-fighting equipment storage areas, fire hydrants, designated fire-fighting water supply signs, or entrances to fire-fighting water tanks that are adjacent to roads.

(9) No parking and stopping zones and times

● It's "Parking" if loading or unloading luggage takes more than **5 minutes**. It's "Stopping" if you stop the vehicles **within 5 minutes**.

● **Within 5 meters** from the intersection and its edge.

● **Within 5 meters** from the road bend.

● **Within 5 meters front and back** from the pedestrian crossing, bicycle crossing, and its edge.

● **Within 10 meters** from the signboard (pillar) of the bus and streetcar stop during operation.

● **Within 10 meters** from the crossing and its edge.

● **Within 10 meters front and back** from the left side of the safety zone and its edge.

（10）路側帯のある道路での駐停車（ろそくたいのあるどうろでのちゅうていしゃ）

一本線に路側帯がある道路

 路側帯の幅が 0.75 m以下の場合は車道の左端に沿って中には入らない。

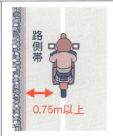 路側帯の幅が 0.75 m以上を超える場合は，中に入って左側に 0.75 m以上の余地をあける。

決められた余地をあける

 車の右側の道路上で 3.5 m以上の余地がない場所では，駐車できない。

 標識で余地が指定されている場所では，車の右側の道路上にその長さ以上の余地をあける。

（11）衝撃力・遠心力・制動距離（しょうげきりょく・えんしんりょく・せいどうきょり）

衝撃力と遠心力，制動距離は速度の2乗に比例する。
衝撃力は速度の2乗に比例するので，高速運転するときは気をつけよう。

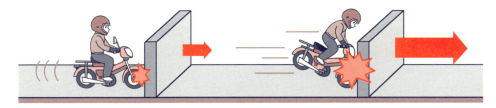

(10) Parking and stopping on roads with side strips.

A road with a side strip

	If the width of side strip is **less than 0.75 meters**, do not enter inside and stop along the left side of the roadway.		If the width of the side strip **exceeds 0.75 meters**, enter inside and leave more than 0.75 meters of space on the left.

Leave the room for sufficient width as specified.

	You cannot park in places on the road to the right of the vehicle where there is not **more than 3.5 meters** of space.		In places where the margin is specified by signs, leave **more than that width** of space on the road to the right of the vehicle.

(11) Impact force, centrifugal force, braking distance.

Impact force and centrifugal force, as well as braking distance, are proportional to **the square** of the speed.

Since impact force is proportional to **the square** of the speed, be careful when driving at high speeds.

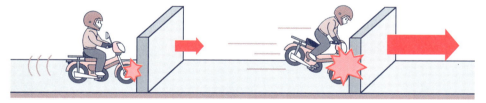

(12) 車の停止距離（くるまのていしきょり）

車はいきなり止まれないのでブレーキをかけてから，車が止まるまでの距離のことを停止距離。
停止距離は，下の空走距離と制動距離を足した距離のことをいう。

空走距離
危険を察知し，ブレーキをかけてから，ブレーキが効き始めるまでに車が走る距離。

(13) 制動距離（せいどうきょり）

実際にブレーキが効き始めてから，車が停止するまでの距離。
空走距離は運転者が疲れているとき（判断までに時間がかかってしまう）は長くなる。
制動距離はタイヤがすり減っていたり，路面が雨で濡れている場合は通常のおよそ2倍になるので注意が必要。

110　5. まちがえやすい数字のまとめ

(12) Stopping distance of a vehicle.

A vehicle cannot stop suddenly, so **the stopping distance** refers to the distance from the moment the brakes are applied until the vehicle comes to a stop.
The stopping distance is the sum of **the reaction distance** and **the braking distance**.

What is the reaction distance?
Reaction distance is the distance a vehicle travels from the moment the driver perceives a danger and applies the brakes until the brakes begin to take effect.

(13) Braking distance.

The distance after the brakes start to take effect until the car comes to a stop.
The reaction distance can become longer when the driver is tired (as it takes more time to make a decision).
Braking distance needs careful attention as it can become approximately **twice** as long as usual if the tires are worn down or if the road surface is wet from rain.

6．イラスト問題の攻略法 （いらすともんだいのこうりゃくほう）

　試験には危険を予測した運転に関するイラスト問題が2問出題されます。どんな危険が潜んでいるか？や，どんな運転行動が安全か？を出題されるので，落ち着いて答えよう。イラスト問題では1問につき3つの設問があり，全て正解して2点です。1つでも間違えると得点にはならないので要注意です。

（1）イラスト問題を解く攻略法！ （いらすともんだいをとくこうりゃくほう）

① 「～するはずなので」「～と思われるので」の表現には要注意！
　　思い込みの判断で運転することは危険。
② 「すばやく」「急いで」などの表現には要注意」！
　　急ブレーキ，急ハンドルは必要性を問われる場合が多い。
③ 「見えないところ」にも要注意！
　　危険はどこにでも潜んでいることを忘れない。
④ 「そのままの速度で」「速度を上げて」などの表現には要注意！
　　徐行，停止などが必要か問われる場合が多い。

112　6．イラスト問題の攻略法

6. Strategies for solving illustration questions.

In the exam, there will be two illustration questions related to driving with anticipation of dangers. Questions will be about what kind of dangers are lurking or what kind of driving actions are safe. So, keep calm to solve the questions. For illustration-based questions, there are three questions per question, and you get 2 points for answering all correctly. Be careful, as you won't score any points if even one is answered incorrectly.

(1) Solving strategy for illustration-based questions!

① Be careful with expressions like **'should do'** or **'is expected to.'**
  Operating on assumptions can be dangerous.
② Watch out for expressions like **'quickly'** or **'in hurry'**!
  Sudden braking or steering is often questioned for its necessity.
③ Pay attention to **'invisible areas'**
  Don't forget that dangers could be lurking anywhere.
④ Be cautious with expressions like **'maintain the current speed'** or **'increase the speed'**!
  Often, it is questioned whether reducing speed or stopping is necessary.

（2）イラスト問題ではここをチェックしよう！

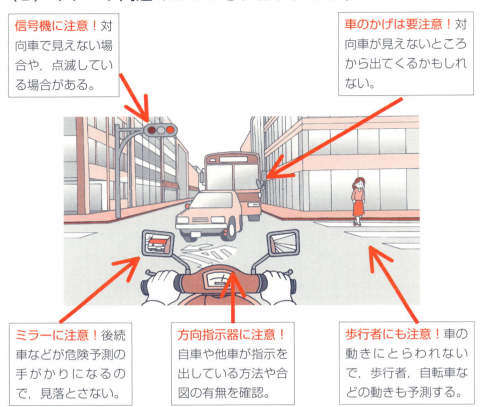

信号機に注意！ 対向車で見えない場合や，点滅している場合がある。

車のかげは要注意！ 対向車が見えないところから出てくるかもしれない。

ミラーに注意！ 後続車などが危険予測の手がかりになるので，見落とさない。

方向指示器に注意！ 自車や他車が指示を出している方法や合図の有無を確認。

歩行者にも注意！ 車の動きにとらわれないで，歩行者，自転車などの動きも予測する。

(2) Check these points in illustration-based questions!

Pay attention to traffic lights! There may be cases where they are not visible due to oncoming traffic, or they might be flashing.

Watch out behind the vehicles! Oncoming vehicles may emerge from spots where they're not visible.

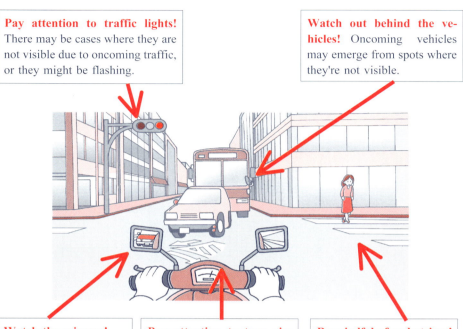

Watch the mirrors! The presence of vehicles behind you, such as those following, can serve as clues for predicting danger, so don't overlook them.

Pay attention to turn signals! Check whether your own vehicle or others are signaling and whether the signals are being used properly.

Be mindful of pedestrians! Don't get too focused on vehicle movements; anticipate the actions of pedestrians, cyclists, and others as well.

7. 危険予測の練習問題 (きけんよそくのれんしゅうもんだい)

問1　時速20キロメートルで進行しています。どのようなことに注意して運転しますか？

(1) トラックの後ろにいる人は，自車の接近に気づいて，道路を横断することはないので，そのままの速度で通行する。
(2) トラックの後ろにいる人は，荷物を運ぶため道路を横断するかもしれないので，いつでも止まれるように速度を落として通行する。
(3) 左側の門から，荷物を取りに人が出てくるかもしれないので，いつでも止まれるように速度を落として通行する。

解答と解説
(1) ×　トラックの後ろの人は，自車の接近に気づくとは限らないです。
(2) ○　トラックの後ろの人の行動に注意しながら，いつでも停車できるように速度を落として通行します。
(3) ○　左側の門にも注意して，安全な速度で通行します。

7. Excercises for predicting dangers.

Question 1: You are traveling at a speed of 20 kilometers per hour. What do you pay attention to while driving

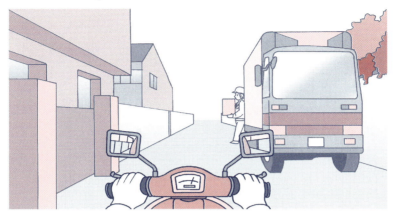

(1) The person behind the truck will notice the approach of your vehicle and will not cross the road, so you continue at the same speed.
(2) The person behind the truck might cross the road to carry luggage, so reduce speed to be able to stop at any time.
(3) Someone might come out from the gate on the left to pick up luggage, so reduce speed to be able to stop at any time.

Answers and explanations.
(1) × The person behind the truck may not notice the approach of your vehicle.
(2) ○ Pay attention to the actions of the person behind the truck and **reduce speed** so that you can stop at any time.
(3) ○ Pay attention to the gate on the left and pass at **a safe speed**.

問2 雨の日に時速20キロメートルで進行しています。どのようなことに注意して運転しますか？

(1) 歩行者は傘をさしていて，自車の接近に気づきにくいので，速度を落として，歩行者の動きに十分注意して通行する。
(2) 子どもがふざけて自車の前に飛び出してくるかもしれないので，速度を落として，子どもに十分注意して通行する。
(3) このまま進行すると歩行者に雨水をはねてしまうおそれがあるので，速度を落として，注意して通行する。

解答と解説
(1) ○　雨の日は，歩行者が傘をさしているので，車の接近に気づかないときがあります。
(2) ○　子どもの動きに十分注意して，速度を落として通行します。
(3) ○　歩行者に対して，雨水がはねないように注意して通行します。

Question 2: You are traveling at a speed of 20 km/h on a rainy day. What do you pay attention to while driving?

(1) Pedestrians holding umbrellas may not notice the approach of your vehicle, so reduce speed and pay close attention to the movements of pedestrians.
(2) Children may playfully get in front of your vehicle, so reduce speed and pay close attention to children.
(3) Continuing at the current speed may splash rainwater on pedestrians, so reduce speed and proceed with caution.

Answer and explanation.
(1) ○   On rainy days, pedestrians using umbrellas may not notice the approach of a car.
(2) ○   Pay close attention to the movements of children and **reduce speed** while passing.
(3) ○   **Be careful** not to splash rainwater on pedestrians while passing.

7. Excercises for predicting dangers.

模擬試験 Practice test.

第1回　模擬試験	First practice test.	(P.122)
解答解説	Answers and explanations.	(P.130)
第2回　模擬試験	Second practice test.	(P.138)
解答解説	Answers and Explanations.	(P.146)
第3回　模擬試験	Third practice test.	(P.154)
解答解説	Answers and Explanations.	(P.162)
第4回　模擬試験	Fourth practice test.	(P.170)
解答解説	Answers and explanations.	(P.178)
第5回　模擬試験	Fifth practice test.	(P.186)
解答解説	Answers and explanations.	(P.194)
第6回　模擬試験	Sixth practice test.	(P.202)
解答解説	Answers and explanations.	(P.210)
第7回　模擬試験	Seventh practice test.	(P.218)
解答解説	Answers and explanations.	(P.226)
第8回　模擬試験	Eighth practice test.	(P.234)
解答解説	Answers and explanations.	(P.242)

第1回 模擬試験	問1〜問46までは各1点，問47，48は全て正解して各2点。制限時間30分，50点中45点以上で合格

●次の問題で正しいものは「正」，誤りのものは「誤」の枠をぬりつぶして答えなさい。

問1 正誤 □□	前の車に続いて踏切を通過するときは，安全を確認すれば一時停止する必要はない。
問2 正誤 □□	車は路側帯の幅の広さにかかわらず，路側帯の中にはいって停車してはならない。
問3 正誤 □□	横断歩道を通過するときは，歩行者がいないときでも一時停止をしなければならない。
問4 正誤 □□	発進する場合は，方向指示器などで合図をして，もう一度バックミラーなどで前後左右の安全を確認するとよい。
問5 正誤 □□	交通事故を起したときは，負傷者の救護より先に警察や会社などに電話で報告しなければならない。
問6 正誤 □□	原動機付自転車が一方通行の道路から右折するときは，道路の左側に寄り，交差点の内側を徐行して通行しなければならない。
問7 正誤 □□	原動機付自転車を運転するときは，免許証に記載されている条件を守らないといけない。
問8 正誤 □□	ブレーキは一度に強くかけないで，数回に分けるとよい。
問9 正誤 □□	右の図の標識があるところでは路面にでこぼこがあるので，注意して運転しなければならない。 
問10 正誤 □□	停止距離とは，空走距離と制動距離を合わせた距離をいう。
問11 正誤 □□	原動機付自転車では，30キログラムまで積むことができる。
問12 正誤 □□	右の図の標識のある交差点で直進する場合は，右側か真ん中の通行帯を通行する。 
問13 正誤 □□	横断歩道の手前で止まっている車があるときは，その車の側方を徐行して通過しなければならない。

122　第1回模擬試験　問題

First Practice Test.	Questions 1 to 46 are worth 1point each, while questions 47 and 48 are worth 2 points each if all answers are correct. The time limit is 30 minutes. To pass, you need 45 points out of 50.

● Fill in the box marked 'T' for the correct answers and 'F' for the incorrect ones in the following questions.

Q1. T F ☐☐	When passing through a railway crossing following the preceding vehicle, it is not necessary to temporarily stop as long as safety is ensured.
Q2. T F ☐☐	Regardless of the width of the roadside belt, vehicles must not enter and stop within the roadside belt.
Q3. T F ☐☐	When crossing a pedestrian crossing, you must come to a stop even if there are no pedestrians.
Q4. T F ☐☐	When starting, it's advisable to signal with indicators and then check the safety of the surroundings using rearview and side mirrors before proceeding.
Q5. T F ☐☐	When involved in a traffic accident, it is necessary to report to the police or the company by phone before providing assistance to the injured.
Q6. T F ☐☐	When a motorized bicycle turns right from a one-way street, it must keep to the left side of the road and proceed slowly on the inside of the intersection.
Q7. T F ☐☐	When driving a motorized bicycle, you must adhere to the conditions listed on the license.
Q8. T F ☐☐	It is advisable to apply the brakes gently in stages rather than applying them strongly all at once.
Q9. T F ☐☐	You must drive carefully in places where there are signs like the one on the right, as there may be uneven road surfaces.
Q10. T F ☐☐	The stopping distance refers to the combined distance of the reaction distance and the braking distance.
Q11. T F ☐☐	In a motorized bicycle, you can load up to 30 kilograms.
Q12. T F ☐☐	When proceeding straight at an intersection with the sign on the right, you should use the right or center lane.
Q13. T F ☐☐	When there is a car stopped before a pedestrian crossing, you must pass it slowly on its side.

First Practice Test. 123

問14 正誤 ☐☐	交差点を通行中に緊急自動車が近づいてきたときは，ただちに交差点の隅に寄って，一時停止をしなければならない。
問15 正誤 ☐☐	原動機付自転車を運転するときは，乗車用ヘルメットをかぶらなければならない。
問16 正誤 ☐☐	他の車に追い越されるときは，できるだけ左側に寄り，その車が追い越し終わるまで，速度を上げてはならない。
問17 正誤 ☐☐	信号が青色でも，前方の交通が混雑しているため交差点の中で動きがとれなくなりそうなときは，交差点に入ってはならない。
問18 正誤 ☐☐	上り坂で停止するとき，前の車に接近しすぎないように止めるとよい。
問19 正誤 ☐☐	右の図の標識がある道路では，自動車は通行できないが，原動機付自転車は通行できる。 
問20 正誤 ☐☐	交通整理が行われていない，道幅が同じような交差点（環状交差点，優先道路通行中の場合は除く）では左方からくる車はあるとき，その車の進行を妨げてはならない。
問21 正誤 ☐☐	上り坂の頂上付近とこう配の急な下り坂は，追越しが禁止されている。
問22 正誤 ☐☐	走行中，アクセルワイヤーが引っ掛かり，アクセルが戻らなくなったら，急ブレーキをかけて止まる。
問23 正誤 ☐☐	右の標識があるところでは，原動機付自転車は軌道敷内を通行できる。 
問24 正誤 ☐☐	安全地帯に歩行者がいるときは，徐行して進むことができる。
問25 正誤 ☐☐	原動機付自転車で3車線以上の車両通行帯のある道路を通行中，信号機のある交差点で二段階右折をした。
問26 正誤 ☐☐	ひとり歩きしている子どものそばを通行するときに1メートルぐらい離れていたので，徐行しないで通行した。
問27 正誤 ☐☐	停車している車のそばを通るときは，急にドアがあいたり，歩行者が車のかげから飛び出してくることがあるので注意が必要である。
問28 正誤 ☐☐	右の標識は二輪の自動車のみ通行できることを示している。 

Q14. T F ☐☐	When an emergency vehicle approaches while passing through an intersection, you must immediately pull over to the corner of the intersection and come to a temporary stop.
Q15. T F ☐☐	When driving a motorized bicycle, you must wear a helmet designed for driving.
Q16. T F ☐☐	When being overtaken by another vehicle, you should move as far left as possible and refrain from increasing speed until the overtaking is completed.
Q17. T F ☐☐	Even if the signal is green, you must not enter the intersection if it seems that you won't be able to move within the intersection due to congestion ahead.
Q18. T F ☐☐	When stopping on an uphill slope, it is advisable to stop without getting too close to the vehicle in front.
Q19. T F ☐☐	In roads with the sign on the right, automobiles cannot pass, but motorized bicycles can pass.
Q20. T F ☐☐	At intersections where traffic control is not being conducted and the road widths are similar (excluding roundabouts and when driving on priority roads), if a vehicle is coming from the left, you must not obstruct its progress.
Q21. T F ☐☐	Overtaking is prohibited near the crest of an uphill slope and on steep downhill slopes.
Q22. T F ☐☐	If the accelerator wire gets stuck while driving and the accelerator does not return, apply the brakes firmly to stop.
Q23. T F ☐☐	In areas with the sign on the right, motorized bicycles are allowed to pass through the railway tracks.
Q24. T F ☐☐	You may proceed slowly when there are pedestrians in the safety zone.
Q25. T F ☐☐	I performed a two-stage right turn at a signalized intersection while driving a motorized bicycle on a road with three or more lanes for traffic.
Q26. T F ☐☐	I passed by a child walking alone while keeping a distance of about 1 meter, without slowing down.
Q27. T F ☐☐	When passing by a stationary car, it is important to be cautious as doors may suddenly open or pedestrians may emerge behind the car.
Q28. T F ☐☐	The sign on the right indicates that only two-wheeled motor vehicles can pass.

First Practice Test.

問29 正誤 ☐☐	初心運転者期間とは，普通免許，大型二輪免許，普通二輪免許，原付免許を取得後1年間のことをいう。
問30 正誤 ☐☐	徐行とは15〜20キロメートル毎時の速度である。
問31 正誤 ☐☐	エンジンをかけた原動機付自転車を押して歩く場合は，歩行者として扱われる。
問32 正誤 ☐☐	運転中，マフラーが故障して大きな排気音を発する状態になったが，運転上危険ではないのでそのまま運転してよい。
問33 正誤 ☐☐	雪道では先に走った車のタイヤの跡を避けて走った方が安全である。
問34 正誤 ☐☐	右の図の標識のある道路で原動機付自転車を右側にはみ出さずに追越しをした。 
問35 正誤 ☐☐	ぬかるみや砂利道を通るときは，トップギアで惰性をつけて通行するとよい。
問36 正誤 ☐☐	車両通行帯のない道路では，中央線から左側ならどの部分を通行してもよい。
問37 正誤 ☐☐	踏切では一時停止をして，自分の目と耳で左右の安全を確かめなければならない。
問38 正誤 ☐☐	歩行者の通行や他の車などの正常な通行を妨げるおそれがあるときは，横断や転回が禁止されていなくても横断や転回をしてはならない。
問39 正誤 ☐☐	霧の中を走るときは，前照灯をつけ，危険防止のため必要に応じて警音器を鳴らすとよい。
問40 正誤 ☐☐	黄色の灯火の点滅は，必ず一時停止をして安全確認をしてから進まなければならない。
問41 正誤 ☐☐	右の標識のある交差点で原動機付自転車が右折する場合は，交差点の側端に沿って徐行する二段階右折をしなければならない。 
問42 正誤 ☐☐	交差点では，左折する車の後輪に巻き込まれるおそれがあるので，車の運転者からよく見える位置を走行するようにしなければならない。
問43 正誤 ☐☐	夜間，繁華街がネオンや街路灯などで明るかったので，原動機付自転車の前照灯をつけないで運転した。

126　第1回模擬試験　問題

Q29. T F ☐☐	The beginner rider period refers to the one-year period after obtaining a standard motor vehicle license, a large-size motorcycle license, a regular motorcycle license, or a moped license.
Q30. T F ☐☐	Driving at reduced speed refers to a speed of 15 to 20 km/h.
Q31. T F ☐☐	When walking with a motorized bicycle with the engine running, you are treated as a pedestrian.
Q32. T F ☐☐	While driving, if the muffler malfunctions and emits a loud exhaust noise, as long as it does not pose a danger while driving, you can continue driving as is.
Q33. T F ☐☐	It is safer to avoid the tire tracks of the vehicle ahead when driving on snowy roads.
Q34. T F ☐☐	I overtook without protruding to the right on a road with the sign shown on the right. 
Q35. T F ☐☐	When passing through mud or gravel roads, it is advisable to use top gear and coast.
Q36. T F ☐☐	On roads without traffic lanes, you may drive on any part to the left of the center line.
Q37. T F ☐☐	At railway crossings, you must come to a temporary stop and verify safety to the left and right using your own eyes and ears.
Q38. T F ☐☐	You must not cross or turn even if crossing or turning is not prohibited when there is a possibility of obstructing pedestrian or other normal traffic.
Q39. T F ☐☐	When driving in foggy conditions, it is advisable to turn on the headlights and, if necessary for safety, sound the horn.
Q40. T F ☐☐	Flashing yellow lights require you to come to a complete stop and verify safety before proceeding.
Q41. T F ☐☐	When a motorized bicycle makes a right turn at an intersection with the sign on the right, it must perform a two-stage right turn by proceeding slowly along the edge of the intersection. 
Q42. T F ☐☐	At intersections, you must drive in a position visible to the driver of the vehicle to avoid the risk of being caught by the rear wheels of a left-turning car.
Q43. T F ☐☐	I drove without turning on the headlight of the motorized bicycle at night because the downtown area was bright with neon lights and streetlights.

First Practice Test. 127

問44 正誤 □□	追越ししようとするときは，その場所が追越し禁止場所ではないかを確かめる。
問45 正誤 □□	ブレーキは道路の摩擦係数が小さくなればなるほど強くかかる。
問46 正誤 □□	同一方向に進行しながら進路変更するときは，進路を変更するときの10秒前に合図をしなければならない。

問47　雨上がりの道路を時速30キロメートルで進行しています。この場合，どのようなことに注意して運転しますか？

(1) 正誤 □□　雨で濡れた道路での急なハンドル操作は，転倒の原因になるので，速度を落として慎重に運転する。

(2) 正誤 □□　雨で濡れた道路での急ブレーキは，横滑りの原因になるので，早めにスロットルを戻し，速度を落とす。

(3) 正誤 □□　雨で濡れた路面のカーブでは，曲がりきれず，中央線をはみ出すおそれがあるので，手前の直線部分で十分に速度を落とす。

問48　夜間，右折待ちのため停止しています。この場合，どのようなことに注意して運転しますか？

(1) 正誤 □□　トラックが右折するときに，トラックの右側を同時に右折すると安全である。

(2) 正誤 □□　トラックの陰から直進してくる対向車があるかもしれないので，トラックが右折したあとに続いて右折せず，安全を確認してから右折する。

(3) 正誤 □□　夜間は車のライトが目立つため，歩行者は自車の存在に気づいて立ち止まるので，両側の歩行者の間を右折する。

Q44. T F ☐☐	When attempting to overtake, it is important to verify whether the location is not designated as a no-passing zone.
Q45. T F ☐☐	The brakes apply more forcefully as the coefficient of friction of the road decreases.
Q46. T F ☐☐	When changing lanes while moving in the same direction, you must signal 10 seconds before changing lanes.

Q47.　When driving at 30 kilometers per hour on a road after rain, what should you pay attention to?

(1)　Sudden steering on wet roads due to rain can cause accidents, so drive carefully and
T F　reduce speed.
☐☐

(2)　Sudden braking on wet roads due to rain can cause skidding, so release the throttle
T F　early and reduce speed.
☐☐

(3)　On wet road surfaces during curves, there is a risk of not being able to make the turn
T F　and crossing the centerline, so slow down sufficiently on the straight section before
☐☐　the curve.

Q48.　What should you pay attention to while waiting to turn right at night?

(1)　It is safe to simultaneously turn right alongside a truck when the truck is turning
T F　right.
☐☐

(2)　You should not immediately follow the truck's right turn and should verify safety
T F　before making the right turn, as there may be oncoming vehicles coming straight
☐☐　from behind the truck.

(3)　At night, pedestrians may notice the presence of a car due to its headlights and stop,
T F　so it is safe to turn right between pedestrians on both sides.
☐☐

第1回 解答と解説	間違えたら赤シートを当てて，覚えておきたいポイントを再チェックしよう！

◆・・・ひっかけ問題　　★・・・重要な問題

問1 誤	踏切を通過するとき，前の車に続いても**一時停止**をし，**安全確認**をしなければいけない。
問2 ★誤	駐停車が禁止されていない幅の広い路側帯の場合には入れる。ただし，道路の端から**0.75メートル**の幅をあけること。
問3 誤	横断している人や横断しようとする人がいるときだけ，横断歩道の手前で**一時停止**する。
問4 ★正	安全確認を怠らないこと。
問5 誤	**負傷者の救護**などを行い，**事故の発生場所**や**負傷者の数**，けがの程度などを警察に報告しなければならない。
問6 ★誤	一方通行の道路では右折するとき，道路の**右側**に寄らなければならない。
問7 正	免許証の記載されている条件を守ること。
問8 正	急ブレーキは危険なので避けること。
問9 正	この標識は「**路面に凹凸あり**」を示しているので，**速度**を落としながら注意して運転する。
問10 正	**空走距離**と**制動距離**を合わせた距離が**停止距離**になる。
問11 正	原動機付自転車の積載物の重量は**30キログラム**までである。
問12 正	この標示は「**進行方向**」を示しているので，直進の場合は**右側**か**真ん中**の通行帯を通行する。
問13 ◆誤	横断歩道の手前で止まっている車の側方を通って前方に出る前に，**一時停止**する。

130　第1回模擬試験　解答と解説

The First Practice Test Answers and Explanations	If you make a mistake, use a red sheet to cover it and re-check the points you want to remember!

◆ · · · Tricky question   ★ · · · Important question

Q1. F	When passing through a railroad crossing, you must come to **a temporary stop** and **perform safety checks** even if following the vehicle in front.
Q2. ★ F	You may enter wide side strips where parking is not prohibited. However, leave a width of **0. 75 meters** from the edge of the road.
Q3. F	**Stop temporarily** just before the pedestrian crossing when there are people crossing or attempting to cross.
Q4. ★ T	Always ensure safety by not neglecting safety checks.
Q5. F	It is necessary to provide **first aid to the injured** and report **the location of the accident, the number of injured persons**, and the extent of injuries to the police.
Q6. ★ F	When making a right turn on a one- way street, you must keep to **the right side** of the road.
Q7. T	Adhere to the requirements stated on your driver's license.
Q8. T	Avoid sudden braking as it is dangerous.
Q9. T	This sign indicates "**uneven road surface**, " so drive carefully while reducing **speed**.
Q10. T	The distance combining **the reaction distance** and **the braking distance** means **the stopping distance**.
Q11. T	The maximum weight for the load of a motorized bicycle is **30 kilograms**.
Q12. T	This marking indicates the "**direction of travel**, " so when proceeding straight, you should use the **right or center** lane for travel.
Q13. ◆ F	Before proceeding past the side of a stopped vehicle near a pedestrian crossing and moving forward, come to **a temporary stop**.

The First Practice Test Answers and Explanations   131

問14 ★誤	緊急自動車が交差点の付近で近づいてきたときは，**交差点を避け**，道路の左側に寄り，**一時停止**する。
問15 ◆正	自動二輪車，原動機付自転車を運転する場合は，**乗車用ヘルメット**をかぶらないといけない。
問16 ★正	追越しの途中に速度をあげると危ないので，**追い越し**が終わるまでできるだけ左側に寄って，速度はあげない。
問17 正	信号が青色でも交差点内で止まってしまいそうなときは，交差点に**入ってはならない。**
問18 正	上り坂では接近しすぎないように気をつける。
問19 誤	この標識は「**通行止め**」を示しているので，この標識のある道路では**歩行者も車も**通行できない。
問20 正	道幅が同じような交差点では，路面電車や左方からくる車の進行を**妨げてはならない。**
問21 ◆正	上り坂の頂上付近やこう配の急な下り坂は追越し禁止場所である。ただし，こう配の急な上り坂は**追越し禁止場所ではない。**
問22 誤	すぐにエンジンスイッチを**切る**などして，エンジンの回転を止める。
問23 ◆誤	この標識は「**軌道敷内通行可**」を示しているので，標識により認められた自動車，右折する場合などは**通行**できる。原動機付自転車は原則として通行できない。
問24 正	歩行者が安全地帯にいるときは，**徐行**しなければいけない。
問25 ★正	車両通行帯が3車線以上ある道路の信号機などがある交差点や，二段階右折の標識がある道路では，**二段階右折**する。
問26 誤	子どもがひとりで歩いているときは，**徐行や一時停止**をして，安全に通行できるようにしなければいけない。
問27 正	止まっている車のそばを通るとき，いきなりドアが開いたり，車のかげから人が飛び出したりする場合があるので注意する。

Q14. ★ F	When an emergency motor vehicle approaches near an intersection, **avoid the intersection**, move to the left side of the road, and come to **a temporary stop**.
Q15. ◆ T	When driving a motorcycle or a motorized bicycle, it is necessary to wear a **helmet designed for riding**.
Q16. ★ T	It's dangerous to increase speed while overtaking, so until **the overtaking** is complete, stay as far left as possible and do not accelerate.
Q17. T	If it seems like you may get stuck in the intersection even when the signal is green, **you should not enter** the intersection.
Q18. T	Be careful not to get too close when approaching an uphill slope.
Q19. F	This sign indicates a "**road closed**". Therefore, **neither pedestrians nor vehicles** are allowed to pass on roads with this sign.
Q20. T	At intersections where the road width is similar, one must **not obstruct** the movement of streetcars or vehicles approaching from the left.
Q21. ◆ T	Overtaking is prohibited near the top of an uphill slope and on steep downhill slopes with sharp curves. However, steep uphill slopes are **not areas where overtaking is prohibited**.
Q22. F	Immediately **turn off** the engine switch to stop the engine from rotating.
Q23. ◆ F	This sign indicates "**permitted passage in the streetcar tracks**," **allowing passage** for vehicles authorized by the sign, such as those making right turns. However, motorized bicycles are generally not permitted to pass.
Q24. T	When pedestrians are in a safety zone, you must drive **at reduced speed**.
Q25. ★ T	At intersections with traffic signals or signs indicating two-stage right turns on roads with three or more lanes for vehicular traffic, perform **a two-stage right turn**.
Q26. F	When a child is walking alone, you must **reduce speed** or come to **a temporary stop** to ensure safe passage.
Q27. T	When passing by a stationary vehicle, be cautious as doors may suddenly open or people may emerge behind the car.

The First Practice Test Answers and Explanations   133

問28 誤	この標識は「自転車専用」を示しているので，歩行者，普通自転車以外の車の通行が禁止されている。
問29 正	普通免許，大型二輪免許，普通二輪免許，原付免許では，各免許の種類ごとに取得後1年間を初心運転期間という。（停止中の期間は除く）
問30 誤	おおむね10キロメートル毎時とされていて，車がすぐに停止できる速度で進行することを徐行という。
問31 誤	エンジンをかけていると歩行者として扱われない。
問32 ★誤	騒音など他人に迷惑を与えるおそれのある車は運転できない。
問33 誤	雪道ではタイヤの跡を走行するほうが安全である。
問34 正	この標識は「追越しのための右側部分はみ出し通行禁止」を表しているので，右側部分にはみ出なければ追越しできる。
問35 誤	ぬかるみや砂利道などは，低速ギアで速度を落として通行する。
問36 誤	追越しなどやむを得ない以外は，道路の左側に寄って通行する。
問37 ◆正	踏切ではその直前で一時停止。その後，左右の安全確認しなければいけない。
問38 ★正	歩行者の通行やほかの車などの正常な通行を妨げるおそれがある場合は，横断や転回をしてはいけない。
問39 ★正	霧の中を走行するときは前照灯をつけて，必要に応じて危険防止のため警音器を使用する。
問40 ★誤	黄色の灯火の点滅の場合はほかの交通に注意しながら通行できる。
問41 ★正	この標識は「原動機付自転車の右折方法（二段階）」を示していて，二段階右折をする。

Q28. F	This sign indicates "**Bicycle Only**". Therefore, other vehicle traffic besides pedestrian and regular bicycles is **prohibited**.
Q29. T	For standard motor vehicle licenses, large motorcycle licenses, standard motorcycle licenses, and motorized bicycle licenses, the first year after obtaining each type of license, excluding periods of suspension, is referred to as **the novice driving period**.
Q30. F	Speeds generally around 10kilometers per hour, allowing vehicles to come to an immediate **stop**, are referred to as "reduce speed" or "crawling speed."
Q31. F	If the engine is running, the person is not considered a **pedestrian**.
Q32. ★ F	Vehicles that may cause noise or other disturbances to others cannot be driven.
Q33. F	It is **safer** to drive within the tracks of tires on snowy roads.
Q34. T	This sign indicates "**prohibition of overtaking by protruding into the right-hand lane for overtaking**". Therefore, you can **overtake** as long as you don't protrude into the right-hand lane.
Q35. F	**Drive at a reduced speed** using low gear when encountering muddy or gravel roads.
Q36. F	Except when overtaking or when unavoidable, drive close to the **left side** of the road.
Q37. ◆ T	At a railroad crossing, come to a temporary stop just before it. Afterward, you must check left and right **for safety** before proceeding.
Q38. ★ T	Do not **cross** or **make** turns if there is a risk of obstructing the passage of pedestrians or other vehicles in normal operation.
Q39. ★ T	When driving in foggy conditions, turn on **the headlights** and use a **warning horn** if necessary for safety precautions.
Q40. ★ F	You can **proceed** with caution while paying attention to other traffic when the yellow light is flashing.
Q41. ★ T	This sign indicates the "**two-stage right turn for motorized bicycles**," requiring **a two-stage right turn**.

The First Practice Test Answers and Explanations   135

問42 正	左折する車の後輪に巻き込まれないよう注意。
問43 誤	夜間に道路を通行するときは，街路灯やネオンで明るくても**前照灯**などをつけなければいけない。
問44 ★正	追越しするときは，その場所が**追越し禁止**ではないことを確認してから行う。
問45 誤	道路の摩擦係数が**大きくなる**ほどブレーキは強くかかる。
問46 ◆誤	合図を行う時期は，進路変更する約**3秒前**である。

問47　**路面**と**天候**の状態に要注目！！
道路に水たまりがあると**滑りやすい**状態で，落ち葉でさらにタイヤが**スリップ**しやすいので，慎重に運転しましょう。

(1)　正　雨や落ち葉などで，**横滑り**するおそれがある。
(2)　正　早めに速度を落として，**慎重**に運転する。
(3)　正　カーブ内でブレーキをかけずにすむよう，手前の直線部分で速度を十分に落とす。

問48　トラックの**陰**と**歩行者**に要注目！！
夜間なので周囲が暗いため，**歩行者**の動きに注意しよう。
トラックの**陰**に対向車が隠れているかもしれません。

(1)　**誤**　トラックの右側を右折すると**歩行者に接触する**おそれがある。
(2)　正　トラックの陰から**直進車**が出てくるおそれがある。
(3)　**誤**　歩行者は自車の存在に**気づかない**おそれがある。

Q42. T	Be careful not to get caught in the rear wheels of a car turning left.
Q43. F	When traveling on roads at night, even if there are bright streetlights or neon signs, you must use **headlights and other lights**.
Q44. ★ T	Before overtaking, ensure that the location is not **prohibited for overtaking**.
Q45. F	**The greater** the coefficient of friction on the road, the stronger **the brakes** will engage.
Q46. ◆ F	The appropriate time to signal is approximately **3 seconds before** changing lanes.

Q47.   Pay attention to the condition of **the road surface** and **the weather!**
Puddles on the road can make it **slippery**, and fallen leaves can further increase the risk of tire **slipping**. Proceed with caution.

(1)   T   There is a risk of **skidding** due to rain or fallen leaves.
(2)   T   Reduce speed in advance and drive **carefully**.
(3)   T   To avoid braking in curves, sufficiently reduce speed on the straight section before the curve.

Q48.   Pay attention to the **blind spots** of trucks and **pedestrians!**
Since it's nighttime and the surroundings are dark, be vigilant about **pedestrians'** movements.
Oncoming vehicles may be hidden **behind** the truck.
(1)   F   There is a risk of **colliding with pedestrians** when turning right on the right side of a truck.
(2)   T   There is a risk of **straight-going vehicles** emerging behind a truck.
(3)   F   There is a risk of pedestrians **not being aware** of the presence of your vehicle.

第2回模擬試験	問1〜問46までは各1点，問47，48は全て正解して各2点。 制限時間30分，50点中45点以上で合格

●次の問題で正しいものは「正」，誤りのものは「誤」の枠をぬりつぶして答えなさい。

問1 正誤 □□	交差点内を通行しているとき，緊急自動車が近づいてきたので，ただちに交差点の中で停止した。
問2 正誤 □□	夜間，見通しの悪い交差点で車の接近を知らせるために，前照灯を点滅した。
問3 正誤 □□	原動機付自転車は，道路が渋滞しているときでも機動性に富んでいるので，車の間をぬって走ることができる。
問4 正誤 □□	交差点以外で，横断歩道も自転車横断帯も踏切もないところに信号機があるときの停止位置とは，信号機の直前である。
問5 正誤 □□	子どもが急に飛び出してきたので，これを避けるために急ブレーキをかけた。
問6 正誤 □□	上り坂の頂上付近では，徐行の標識がなくても，常に徐行しなければならない。
問7 正誤 □□	原動機付自転車は，路線バスの専用通行帯を通行することができるが，その場合，バスの通行を妨げないようにしなければいけない。
問8 正誤 □□	チェーンのゆるみ具合は，車に乗った状態で点検する。
問9 正誤 □□	右の図の標識のある場所では，停止線の直前で一時停止するとともに，交差する道路を通行する車の通行を妨げてはいけない。 
問10 正誤 □□	消火栓や防火水そうなどの消防施設のあるところから5メートル以内には，原動機付自転車を駐車してはならない。
問11 正誤 □□	車がカーブを曲がるとき，車が外側に飛び出そうとするのは，車の重心が移動するからである。
問12 正誤 □□	右の標識がある道路では，前方に「道路工事中」のところがあることを表している。 
問13 正誤 □□	信号機の信号が赤色の点滅を標示しているときは，一時停止をし，安全確認をした後に進行することができる。

Second Practice Test.	Questions 1 to 46 are worth 1point each, while questions 47 and 48 are worth 2 points each if all answers are correct. The time limit is 30 minutes. To pass, you need 45 points out of 50.

● Fill in the box marked 'T' for the correct answers and 'F' for the incorrect ones in the following questions.

Q1. T F ☐☐	When driving through an intersection, I immediately stopped in the intersection when an emergency vehicle approached.
Q2. T F ☐☐	At night, at poorly visible intersections, I flashed the headlights to indicate the approach of a vehicle.
Q3. T F ☐☐	Motorized bicycles are highly maneuverable, so even in heavy traffic, they can navigate between cars.
Q4. T F ☐☐	When there are no intersections, crosswalks, or railway crossings, and there is a traffic signal, the stopping position is just before the traffic signal.
Q5. T F ☐☐	I applied the brakes suddenly to avoid a child who suddenly darted out.
Q6. T F ☐☐	Near the top of an uphill slope, even if there are no signs indicating "reduce speed, " you must always drive at reduced speed.
Q7. T F ☐☐	Motorized bicycles are allowed to use lanes designated for bus use, but in doing so, they must not impede the operation of buses.
Q8. T F ☐☐	The looseness of the chain should be inspected while the rider is on the bike.
Q9. T F ☐☐	At the location marked with the sign in the diagram on the right, you must come to a temporary stop just before the stop line and must not obstruct the passage of vehicles traveling on the intersecting road.
Q10. T F ☐☐	Motorized bicycles must not be parked within 5 meters of firefighting facilities such as fire hydrants or fire water tanks.
Q11. T F ☐☐	When a car is turning a curve and appears to be veering outward, it is because the car's center of gravity is shifting.
Q12. T F ☐☐	The sign on the right indicates that there is "road works" on the road.
Q13. T F ☐☐	When the traffic signal displays flashing red lights, you may proceed after coming to a temporary stop and ensuring it is safe to do so.

Second Practice Test. 139

問14 正誤	横断歩道の手前から 30 メートル以内は，追越しは禁止されているが，追抜きはよい。
問15 正誤	交通整理が行われていない道幅が同じような交差点に，左右から同時に車がさしかかったときは，左方車が右方車に優先する。
問16 正誤	「車両横断禁止」の標識があったが，道路の左側にある車庫に入るため左側に横断した。
問17 正誤	道路の正面に障害物があったが，対向車より先に障害のある場所に到達したので，急いで通過した。
問18 正誤	眠気をもよおす風邪薬を飲んだ時は，運転をひかえるようにする。
問19 正誤	右の図の標識のある道路では，二輪の自動車は通行できないが，原動機付自転車は通行できる。 
問20 正誤	原動機付自転車に乗る人は，大型自動車の死角や内輪差を知っていた方がよい。
問21 正誤	車から離れるときは，原動機付自転車が倒れないようにスタンドを立て，必ずハンドルロックをしてキーを抜くようにする。
問22 正誤	原動機付自転車に同乗する人も，つとめてヘルメットをかぶらなければいけない。
問23 正誤	右の標識のある通行帯を原動機付自転車で通行中に路線バスが接近してきたときは，その通行帯から出なければならない。 
問24 正誤	幅が 0.75 メートルを超える白線 1 本の路側帯のある場所で駐停車するときは，路側帯の中に入り，車の左側に 0.75 メートル以上の余地を残す。
問25 正誤	止まっている通学・通園バスのそばを通るとき，保護者が児童に付き添っていたので，徐行しないでその側方を通過した。
問26 正誤	警笛区間の標識がある区間内にある交差点を通過するときは，どんな場合でも警音器を鳴らさなければいけない。
問27 正誤	信号機の黄色の矢印信号に対面した原動機付自転車は，停止位置から先に進むことができない。
問28 正誤	右の図の標識は，この先にゆるやかな上り坂があることを表している。 

140　第 2 回模擬試験　問題

Q14. T F □□	Overtaking is prohibited within 30 meters before a pedestrian crossing, but passing is allowed.
Q15. T F □□	When two vehicles approach an intersection with similar road widths and no traffic control, the vehicle on the left yields to the vehicle on the right.
Q16. T F □□	There was a "No Right Turn Crossing Ahead" sign, but I crossed to the left to enter a garage located on the left side of the road.
Q17. T F □□	There was an obstacle ahead on the road, but I hurriedly passed through because I reached the obstructed area before the oncoming vehicle did.
Q18. T F □□	When taking cold medicine that causes drowsiness, it is important to refrain from driving.
Q19. T F □□	On the road with the sign in the right diagram, motorcycles are not allowed to pass, but motorized bicycles are allowed to pass.
Q20. T F □□	Riders of motorized bicycles should be aware of blind spots and turning radius of large vehicles.
Q21. T F □□	When moving away from the bike, ensure that the stand is deployed to prevent the motorized bicycle from tipping over, and always engage the handle lock and remove the key.
Q22. T F □□	Anyone riding as a passenger on a motorized bicycle must also wear a helmet as a precautionary measure.
Q23. T F □□	When riding a motorized bicycle in a lane with the sign on the right, if a bus approaches, you must exit that lane.
Q24. T F □□	When parking in an area with a single white line roadside with a width exceeding 0. 75 meters, you should park in the side strip, leaving at least 0. 75 meters of space to the left of the vehicle.
Q25. T F □□	When passing by a stopped school or nursery bus where parents were accompanying children, I passed by without reducing speed.
Q26. T F □□	When passing through an intersection in a designated horn zone, you must sound the horn regardless of the situation.
Q27. T F □□	A motorized bicycle facing a yellow arrow signal at a traffic light cannot proceed beyond the stopping position.
Q28. T F □□	The sign in the right diagram indicates that there is a gentle incline ahead.

Second Practice Test. 141

問29 正誤 □□	原動機付自転車は，身体で安定を保ちながら走るという点では，四輪車より運転は難しいといえる。
問30 正誤 □□	速度と燃料消費には密接な関係があり，速度が遅すぎても速すぎても燃料の消費量は多くなる。
問31 正誤 □□	下り坂では，速度が速くなりやすく停止距離が長くなるので，車間距離を長めにとったほうがよい。
問32 正誤 □□	原動機付自転車を運転中に大地震が発生したときは，急ハンドルや急ブレーキを避け，できるだけ安全な方法により道路の左側に停止する。
問33 正誤 □□	雪道や凍結した道路では，低速で速度を一定に保って進行する。
問34 正誤 □□	信号待ちのため一時停止をする場合，右の標示がある部分に入って停止することができる。
問35 正誤 □□	運転免許証を紛失したまま運転していると，無免許運転として処罰される。
問36 正誤 □□	踏切を通過するとき，歩行者や対向車に注意しながら，落輪しないように踏切のやや中央寄りを注意して通行した。
問37 正誤 □□	歩道や路側帯のない場所で，道路外の施設に入るため左折しようとするときは，あらかじめ道路の左端に寄って徐行しなければならない。
問38 正誤 □□	同一方向に2つの車両通行帯がある道路では，高速車は中央寄りの通行帯を，低速車は左側寄りの通行帯を通行する。
問39 正誤 □□	合図は，その行為が終わるまで続け，またその行為が終わったらただちにやめなければならない。
問40 正誤 □□	原動付自転車に積むことのできる積載物の重量は，60キログラムまでである。
問41 正誤 □□	右の標識のあるところでは，原動機付自転車は通行できる。
問42 正誤 □□	無段変速装置付のオートマチック二輪車のスロットルを完全に戻すと，車輪にエンジンの力が伝わらなくなり，安定を失うことがある。
問43 正誤 □□	道路の曲がり角付近を通行するときは，徐行しなければならない。

Q29. T F ☐☐	When riding a motorized bicycle, it can be said that driving is more difficult than a four-wheeled vehicle because it requires maintaining stability with the body while riding.
Q30. T F ☐☐	There is a close relationship between speed and fuel consumption; fuel consumption increases both when the speed is too slow and when it is too fast.
Q31. T F ☐☐	On downhill slopes, speeds tend to increase, and stopping distances become longer, so it is advisable to maintain a longer following distance.
Q32. T F ☐☐	When driving a motorized bicycle and a major earthquake occurs, it is important to avoid sudden steering or braking. Instead, try to stop on the left side of the road using the safest method possible.
Q33. T F ☐☐	On snowy or frozen roads, maintain a steady speed at a low speed.
Q34. T F ☐☐	When coming to a temporary stop at a traffic signal, you can enter and stop in the area indicated on the right.
Q35. T F ☐☐	Driving without a license when your driver's license is lost results in punishment for driving without a license.
Q36. T F ☐☐	When crossing a railroad crossing, I proceeded cautiously, paying attention to pedestrians and oncoming vehicles, and made sure not to cross the tracks so as not to get stuck in the middle of the crossing.
Q37. T F ☐☐	When attempting to turn left to enter a facility outside the roadway in a location without sidewalks or side strips, you must first reduce speed and move to the left edge of the road.
Q38. T F ☐☐	On a road with two lanes in the same direction, faster vehicles should use the lane closer to the center, while slower vehicles should use the lane closer to the left.
Q39. T F ☐☐	Signals must be maintained until the action is completed, and once the action is completed, they must be immediately stopped.
Q40. T F ☐☐	The maximum weight of the load that can be carried on a motorized bicycle is 60 kilograms.
Q41. T F ☐☐	At locations with the sign on the right, motorized bicycles are permitted to pass.
Q42. T F ☐☐	When fully releasing the throttle of an automatic two-wheeled vehicle equipped with a continuously variable transmission, the engine's power may not be transmitted to the wheels, leading to instability.
Q43. T F ☐☐	When passing near a corner on the road, you must drive at reduced speed.

問44 正誤 □□	横断歩道に近づいたときは，横断する人がいないことが明らかな場合のほかは，その手前で停止できるように減速して進まなければならない。
問45 正誤 □□	原動機付自転車で走行中，黄色の杖を持っている歩行者がいるときは，必ず警音器を鳴らさなければならない。
問46 正誤 □□	軽い交通事故を起こしたが，急用があるので，被害者に名前と住所を告げて，用事を済ますために運転を続けた。

問47　30km/h で進行しています。どのようなことに注意して運転しますか？

(1) 正誤 □□　トラックのドアが開いても安全な間隔をあけて，いつでも止まれるような速度で接近し，横断歩道の手前で一時停止する。

(2) 正誤 □□　トラックの前方にある横断歩道を横断している歩行者がいるので，横断歩道の手前で一時停止する。

(3) 正誤 □□　トラックの前方にある横断歩道を歩行者が渡り始めているので，速度を上げて急いで走行する。

問48　右折のため交差点で停止しています。対向車が左折の合図を出しながら交差点に近づいてきたとき，どのようなことに注意して運転しますか？

(1) 正誤 □□　左折の合図をしている対向車が交差点に接近してきているので，対向車を先に左折させてから安全を確認し，右折する。

(2) 正誤 □□　対向車の後方に他の車が見えなかったので，左折の合図をしている対向車より先に，そのまま右折を始める。

(3) 正誤 □□　左折する対向車は歩行者が横断しているため，横断歩道の手前で停止すると考えられるので，対向車が横断歩道を通過する前に右折する。

Q44. T F ☐☐	When approaching a pedestrian crossing, you must reduce speed and be prepared to stop before it unless it is evident that no one is crossing.
Q45. T F ☐☐	When riding a motorized bicycle and encountering a pedestrian carrying a yellow cane, you must always sound the horn.
Q46. T F ☐☐	I caused a minor traffic accident, but I had urgent business, so I told the victim my name and address and continued driving to take care of my errands.

Q47.   You are driving at 30 km/h. What should you pay attention to while driving?

(1) T F ☐☐   You should approach the truck with a safe distance, ensuring you can stop at any time, even if the door opens. And you should come to a temporary stop before the pedestrian crossing.

(2) T F ☐☐   You will temporarily stop before the pedestrian crossing because there are pedestrians crossing the pedestrian crossing in front of the truck.

(3) T F ☐☐   You will increase speed and hurry to pass because pedestrians have begun crossing the pedestrian crossing in front of the truck.

Q48.   You are stopped at the intersection for a right turn. When an oncoming vehicle approaches with a left turn signal, you will pay attention to the following while driving:

(1) T F ☐☐   An oncoming vehicle is approaching the intersection with a left turn signal, so you will let the oncoming vehicle make the left turn first, then you will ensure safety and make the right turn.

(2) T F ☐☐   Since no other vehicles were visible behind the oncoming vehicle, you will begin the right turn before the oncoming vehicle signaling a left turn.

(3) T F ☐☐   The oncoming vehicle is making a left turn, likely stopping before the pedestrian crossing. Therefore, you will make the right turn before the oncoming vehicle passes through the pedestrian crossing.

第2回 解答と解説	間違えたら赤シートを当てて，覚えておきたいポイントを再チェックしよう！

◆・・・ひっかけ問題　　★・・・重要な問題

問1 誤	交差点内で停止するのではなく，交差点を避けて道路の**左側**に寄り，**一時停止**する。
問2 正	カーブや見通しの悪い交差点の手前では，前照灯を**点滅**するか**上向き**に切り替える。
問3 誤	車の間をぬって走ったり，ジグザグ運転は**危険運転**である。
問4 正	何もないところに信号機があるときは，信号機の**直前**が**停止位置**である。
問5 正	危険防止のためにやむを得ない場合は，急ブレーキで**回避**する。
問6 ★正	標識がなくても，上り坂の頂上付近は**徐行場所**に指定されている。
問7 正	バスの通行を妨げないように**通行**する。
問8 誤	**乗車**せずに，チェーンのゆるみ具合の点検を行う。
問9 ★正	停止線の直前で**一時停止**し，交差する道路の通行を妨げない。
問10 ★正	消防施設のあるところから**5メートル**以内は，駐車してはならない。
問11 誤	車がカーブの外側に飛び出そうとするのは，曲がろうとする外側に**遠心力**が働くため。
問12 正	「**道路工事中**」の標識で，この先の道路が工事中であることを示す標識。
問13 正	信号が赤色の点滅のときは，必ず**一時停止**して安全確認をしてから通行する。

The Second Practice Test Answers and Explanations	If you make a mistake, use a red sheet to cover it and re-check the points you want to remember!

◆ · · · Tricky question   ★ · · · Important question

Q1. F	Rather than stopping inside the intersection, You should avoid the intersection and move to **the left side** of the road, then come to **a temporary stop**.
Q2. T	At curves or intersections with poor visibility, **dim** or **switch** the headlights to high beam.
Q3. F	Passing through gaps between cars or driving in a zigzag manner is considered **dangerous driving**.
Q4. T	When there is a traffic signal in a place where there is nothing else, the stop line **before** the traffic signal is considered **the stopping position**.
Q5. T	In unavoidable situations for safety reasons, use emergency braking to **avoid danger**.
Q6. ★ T	Even without a sign, the top of an uphill slope is designated as **a place to proceed with caution**.
Q7. T	**Pass through** without obstructing the bus's operation.
Q8. F	Check the slackness of the chain without **riding the vehicle**.
Q9. ★ T	Come to **a temporary stop** just before the stop line and do not obstruct the traffic on the intersecting road.
Q10. ★ T	You must not park within **5meters** of fire-fighting facilities.
Q11. F	Cars tend to veer outward on curves because **centrifugal force** acts outward on the outside of the curve.
Q12. T	A sign indicating "**Road Work Ahead**," indicating that the road ahead is under construction.
Q13. T	When the signal is flashing red, always **come to a stop**, ensure it's safe, and then proceed.

The Second Practice Test Answers and Explanations   **147**

問14 ★誤	横断歩道とその手前から**30 メートル**以内の場所では，追越しや追抜きは禁止。
問15 正	交通整理が行われていない道幅が同じような交差点では，右方の車は左方から来る車の進行を**妨げてはならない**。
問16 正	「**車両横断禁止**」は右折を伴う横断を**禁止**している標識。
問17 誤	進路の正面に障害物がある場合，その前に**一時停止**か**減速**して，反対方向の車に道を譲る。
問18 正	催眠作用を催す薬など飲んだときは，運転を**控える**。
問19 ★誤	「**二輪の自動車，原動機付自転車通行止め**」の標識なので，自動二輪車と原動機付自転車は通行できない。
問20 正	運転知識を身につけていた方が良い。
問21 正	ハンドルロックをした方が安全である。
問22 ◆誤	原動機付自転車は**二人乗り禁止**である。
問23 ★誤	「**路線バス等優先通行帯**」の標識で，この通行帯を通行している原動機付自転車は左端に寄って，路線バスに進路を譲る。
問24 正	中に入って**駐停車**する。2 本線の路側帯では，車道の左側に沿って**駐停車**する。
問25 誤	保護者が児童を付き添ってるのにかかわらず，通学・通園バスのそばを通るときは，必ず**徐行**する。
問26 誤	区間内の交差点で**見通しのきかない交差点**を通行するときに，警音器を鳴らす。
問27 ★正	黄色の矢印信号は**路面電車専用**の信号である。

Q14. ★ F	Overtaking or passing is prohibited on or within **30 meters** before a pedestrian crossing and its vicinity.
Q15. T	At intersections where traffic control is not being conducted and the road widths are similar, vehicles on the right should **not obstruct** the progress of vehicles approaching from the left.
Q16. T	The " **No Right Turn Crossing Ahead**" sign **prohibits** crossing with a right turn.
Q17. F	When there is an obstacle ahead in the path, **reduce speed** and come to **a temporary stop** before it, yielding to oncoming traffic.
Q18. T	**Refrain** from driving when taking medication that causes drowsiness.
Q19. ★ F	The sign indicates "**No Entry for Motorcycles and Motorized Bicycles**". Therefore, motorcycles and motorized bicycles are not permitted to pass.
Q20. T	It is better to acquire driving knowledge.
Q21. T	It is safer to use a handle lock.
Q22. ◆ F	**Riding double** on a motorized bicycle is **prohibited**.
Q23. ★ F	The sign "**Priority Lane for Fixed-Route Buses**" indicates that motorized bicycles traveling in this lane should keep to the left and yield to buses.
Q24. T	Pull into and **park** in the designated area. On a two-lane roadside, park along the left side of the road.
Q25. F	Regardless of whether parents accompany the child, always drive **at reduced speed** when passing by a school or nursery bus.
Q26. F	Sound the horn when passing through an intersection in the zone where **visibility is poor**.
Q27. ★ T	The yellow arrow signal indicates **a streetcar-exclusive** signal.

The Second Practice Test Answers and Explanations 149

問28 ◆誤	「上り急こう配あり」を表す警戒標識のため，こう配率がおおむね **10%** 以上の傾斜の坂をいう。
問29 正	身体で安定を保ちながら運転する。
問30 ◆正	速度が遅すぎても速すぎても，燃料の消費量は多くなる。
問31 正	下り坂では，停止距離が長くなるため車間距離を長めにとる。
問32 正	できる限り，安全な方法で道路の左側に停止する。
問33 正	雪道や凍結した道路では，低速で慎重に運転する。
問34 ◆誤	「**停止禁止部分**」のため，この中では停止禁止。
問35 誤	免許証不携帯の違反行為になりますが，**無免許運転**にはならない。
問36 正	踏切の**左端**に寄って通行すると，落輪のおそれがあるため，歩行者などに注意しながらやや**中央寄り**を通行する。
問37 正	左折する場合は，あらかじめ道路の左端に寄り**徐行**する。
問38 誤	高速車や低速車の決まりはない。追い越しなどの場合を除き，左側の**通行帯**を通行する。
問39 正	他の車の迷惑になるので，進路変更などが終われば**合図**をやめる。
問40 ★誤	原動機付自転車の積載物の重量は **30 キログラム**までである。
問41 ★誤	「**車両通行止め**」を表す標識で，自動車や原動機付自転車は通行できない。

Q28. ◆ F	The caution sign indicating "**Steep Incline Ahead**" refers to a slope with a gradient generally exceeding **10%**.
Q29. T	Maintain stability while driving.
Q30. ◆ T	The fuel consumption increases regardless of whether the speed is too slow or too fast.
Q31. T	On downhill slopes, it's important to maintain a longer following distance due to the increased stopping distance.
Q32. T	Stop on the left side of the road using the safest method possible.
Q33. T	Drive at reduced speed and carefully on snowy or icy roads.
Q34. ◆ F	"**No Stopping Zone**" means no stopping in this area.
Q35. F	**Not carrying your license** is a violation, but it does not constitute driving without a license.
Q36. T	Passing close to **the left edge** of a level crossing poses a risk of derailment, so it's safer to pass slightly toward **the center** while watching out for pedestrians.
Q37. T	When making a left turn, it is advisable to approach the left edge of the road beforehand and drive **at reduced speed**.
Q38. F	There are no specific rules for fast or slow vehicles. Except for overtaking, vehicles generally travel **in the left lane**.
Q39. T	Stop **signaling** once the lane changes or maneuver is completed to avoid inconveniencing other vehicles.
Q40. ★ F	The maximum weight of the cargo for a motorized bicycle is **30 kilograms**.
Q41. ★ F	The sign indicates "**Road Closed**" meaning that automobiles and mopeds cannot pass through.

問42 正	無段変速装置付のオートマチック二輪車は，エンジンの回転数が低いときには，車輪にエンジンの力が伝わりにくくなる。
問43 正	道路の曲がり角付近では，**徐行**する。
問44 ★正	横断歩道では**速度を落として**，安全運転で通行する。
問45 ★誤	警音器は鳴らさない。**一時停止**や**徐行**で，黄色のつえを持っている歩行者の通行を妨げない。
問46 誤	交通事故を起こしたら，負傷者がいるいないにかかわらず，警察官に**報告**する。

問47　**死角**や**歩行者**に要注目！！
横断歩道を横断している歩行者や，トラックのかげに注意しよう。

(1)　正　安全な間隔をあけて，速度をおとして**死角部分**に注意する。
(2)　正　横断歩道を渡ろうとしている歩行者がいるので，**一時停止**して安全を確かめる。
(3)　誤　駐車車両の側方を通って前方に出るときに**一時停止**をし，安全確認してから進む。

問48　**対向車**に要注目！！
左折する対向車がいるので，急がずに対向車を先に左折させてから，**安全確認**をして右折しよう。

(1)　正　急がずに対向車を先に左折させてから，**安全確認**をして右折する。
(2)　誤　交差点を右折するとき，左折の**合図**をしている対向車がいるときは，対向車を先に行かせるか，自車が先に右折するかを対向車の交差点までの**距離**と**速度**などから判断する。
(3)　誤　対向車が横断歩道の手前で**一時停止**しようとしているときには，対向車の進路を妨げるような右折はしない。

152　第2回模擬試験　解答と解説

Q42. T	Automatic motorcycles with continuously variable transmissions may not efficiently transfer engine power to the wheels when the engine's RPM is low.
Q43. T	Around bends in the road, it is important to **reduce speed**.
Q44. ★ T	**Reduce speed** and drive safely when approaching a pedestrian crossing.
Q45. ★ F	Do not sound the horn. Come to **a stop** or **reduce speed** and do not obstruct the passage of pedestrians holding yellow canes.
Q46. F	If you are involved in a traffic accident, regardless of whether there are injuries or not, you **must report** it to the police.

Q47.　Watch out for **blind spots** and **pedestrians!**
Be cautious of pedestrians crossing the crosswalk and of trucks' blind spots.

(1)　T　Reduce speed and maintain a safe distance to watch out for **blind spots**.
(2)　T　There are pedestrians about to cross the pedestrian crossing, so come to **a temporary stop** and ensure safety.
(3)　F　When moving forward through the side of parked vehicles, come to **a temporary stop**, check for safety, and then proceed.

Q48.　Watch out for **oncoming vehicles!!**
There is an oncoming vehicle preparing to turn left, so let the oncoming vehicle turn left first, then **check for safety** before making a right turn.

(1)　T　Let the oncoming vehicle turn left first without rushing, then **check for safety** before making a right turn.
(2)　F　When turning right at an intersection and there is an oncoming vehicle **signaling** a left turn, decide whether to let the oncoming vehicle go first or make the right turn first based on factors such as **the distance** and **speed** of the oncoming vehicle to the intersection.
(3)　F　If an oncoming vehicle is attempting to come to **a stop** before a pedestrian crossing, avoid making a right turn that would obstruct the oncoming vehicle's path.

第3回 模擬試験	問1〜問46までは各1点，問47，48は全て正解して各2点。 制限時間30分，50点中45点以上で合格

●次の問題で正しいものは「正」，誤りのものは「誤」の枠をぬりつぶして答えなさい。

問1
正誤
ミニカーは50ccであっても，運転するときは普通免許が必要である。

問2
正誤
交通巡視員が信号機の信号と違う手信号をしていたが，交通巡視員の手信号に従わず，信号機の信号に従って通行した。

問3
正誤
走行中に携帯電話を使用すると危険なので，運転する前に電源を切ったり，ドライブモードに設定しておくようにする。

問4
正誤
徐行とは10〜20キロメートル毎時の速度である。

問5
正誤
道路を安全に通行するためには，交通規制を守れば十分であり，互いに相手のことを考えると円滑な交通を阻害するので，相手の立場を考えない。

問6
正誤
警察官が腕を垂直に上げているときは，警察官の身体の正面に対面する交通については，信号機の赤色の灯火と同じ意味である。

問7
正誤
衝撃力は，速度を2分の1に落とすと2分の1になる。

問8
正誤
原動機付自転車でリヤカーなどをけん引する場合の法定最高速度は，時速20キロメートルである。

問9
正誤
右の標識は前方に横断歩道があることを表している。

問10
正誤
明るさが急に変わると，視力は一時的に急激に低下するので，トンネルに入る場合は，その直前に何回も目を閉じたり開いたりしたほうがよい。

問11
正誤
原動機付自転車の積み荷の高さの制限は，地上から2メートル以下である。

問12
正誤
右の標示があるところで原動機付自転車で停止するときは，二輪と表示してある停止線の手前で停止する。

問13
正誤
保険標章の色と数字は，強制保険が満了する年月を表している。

154　第3回模擬試験　問題

Third Practice Test	Questions 1 to 46 are worth 1 point each, while questions 47 and 48 are worth 2 points each if all answers are correct. The time limit is 30 minutes. To pass, you need 45 points out of 50.

● Fill in the box marked 'T' for the correct answers and 'F' for the incorrect ones in the following questions.

Q1. T F	Even for a 50 cc minicar, a regular driver's license is required when driving.
Q2. T F	The traffic warden was signaling a different hand signal from the traffic light, but I followed the signal from the traffic light instead of obeying the hand signal of the traffic warden.
Q3. T F	Using a mobile phone while driving is dangerous, so it's important to turn off the power or set it to drive mode before driving.
Q4. T F	Driving at reduced speed means driving at a speed of 10 to 20 kilometers per hour.
Q5. T F	To ensure safe passage on the road, it is sufficient to obey traffic regulations. You must not think of each other because it interferes with smooth traffic.
Q6. T F	When a police officer's arm is raised vertically, it has the same meaning as a red traffic light for oncoming traffic facing the officer's front.
Q7. T F	The force of impact is halved when the speed is reduced by half.
Q8. T F	The maximum statutory speed limit for towing a trailer with a motorized bicycle is 20 kilometers per hour.
Q9. T F	The sign on the right indicates the presence of a pedestrian crossing ahead.
Q10. T F	When brightness changes suddenly, vision temporarily decreases sharply, so it is advisable to close and open your eyes several times just before entering a tunnel.
Q11. T F	The height limit for the load of a moped is 2 meters or less from the ground.
Q12. T F	When stopping at a location with the sign on the right while riding a motorized bicycle, stop just before the stop line marked with "二輪" (meaning "two-wheeled vehicles").
Q13. T F	The color and number on the insurance sticker indicate the expiration date of the compulsory insurance.

Third Practice Test 155

問14 正誤 □□	環状交差点を左折，右折，直進，転回しようとするときは，あらかじめできるだけ道路の左端に寄り，環状交差点の側端に沿って徐行しながら通行する。
問15 正誤 □□	深い水たまりを通ると，ブレーキ装置に水が入って一時的にブレーキのききがよくなることがある。
問16 正誤 □□	道路工事の区域の端から5メートル以内のところは駐車も停車も禁止されている。
問17 正誤 □□	霧の中を走るときは，前照灯をつけ，危険防止のため必要に応じて警音器を鳴らすとよい。
問18 正誤 □□	黄色の線の車両通行帯のある道路を通行しているときに，緊急自動車が近づいてきても，進路を譲らなくてもよい。
問19 正誤 □□	道路に車を止めて車から離れるときは，危険防止ばかりでなく，盗難防止の措置もとらなければならない。
問20 正誤 □□	前の車が右折のため右側に進路を変えようとしているときは，その左側を通行して追越しをしてもよい。
問21 正誤 □□	バス専用通行帯であっても，小型特殊自動車や原動機付自転車は通行することができる。
問22 正誤 □□	右の標識がある交差点では，直進と左折はできるが右折はできない。
問23 正誤 □□	原動機付自転車を夜間運転するときは，反射性の衣服や反射材のついた乗車用ヘルメットを着用するとよい。
問24 正誤 □□	対向車がセンターラインをはみ出してきたので，やむを得ず初心者マークをつけた車の前に割り込みをした。
問25 正誤 □□	追越しをする場合に限り，最高速度を超えても構わない。
問26 正誤 □□	ブレーキを数回に分けて踏むと制動灯が点滅するので，後続車への合図にもなり，追突防止に役立つ。
問27 正誤 □□	右の標示があるところでは，駐停車が禁止されているところである。　　　　　　　　←黄色
問28 正誤 □□	交差点で警察官が手信号や灯火による信号をしていても，信号機の信号が優先するので，信号機の信号に従わなければならない。

Q14. T F □□	When turning left, right, going straight, or making a U-turn at a roundabout, it is advisable to approach as close to the left edge of the road as possible and proceed along the side of the roundabout while reducing speed.
Q15. T F □□	When passing through deep puddles, water may enter the braking system temporarily improving brake performance.
Q16. T F □□	Parking and stopping are prohibited within 5 meters from the edge of the roadwork zone.
Q17. T F □□	When driving in foggy conditions, it is advisable to turn on the headlights and, if necessary for safety, sound the horn.
Q18. T F □□	Even when driving on a road with a yellow-lined vehicle lane, you do not need to yield your path to emergency vehicles approaching.
Q19. T F □□	When parking the vehicle and leaving it unattended on the road, it is necessary not only for safety but also for theft prevention measures to be taken.
Q20. T F □□	When the preceding vehicle is attempting to change its course to the right for a right turn, it is permissible to overtake and pass it on the left side.
Q21. T F □□	Even in bus lanes, small special vehicles and motorized bicycles are permitted to pass.
Q22. T F □□	At intersections with the indicated sign, you can go straight or turn left, but you cannot turn right.
Q23. T F □□	When driving a motorized bicycle at night, it is advisable to wear reflective clothing or a helmet with reflective materials.
Q24. T F □□	I cut in front of a car with a beginner mark because the oncoming car was encroaching over the center line and I had no choice.
Q25. T F □□	You may exceed the maximum speed limit only when overtaking.
Q26. T F □□	By applying the brakes several times intermittently, the brake lights flash, serving as a signal to following vehicles and helping to prevent rear-end collisions.
Q27. T F □□	Parking or stopping is prohibited in areas where the indicated markings is displayed. Yellow
Q28. T F □□	Even if a police officer is giving hand signals or using lights to direct traffic at an intersection, you must follow the signals of the traffic lights, as they take precedence.

問29 正誤 ☐☐	駐車禁止場所では，たとえわずかな間でも，人待ちのために車を止めてはならない。
問30 正誤 ☐☐	タイヤがパンクしたときは，ただちに急ブレーキをかけて止める。
問31 正誤 ☐☐	ブレーキを強くかけると，短い距離で止まる。
問32 正誤 ☐☐	追越しをしようとするときは，標識や標示により，その場所が追越し禁止場所でないかを確かめる。
問33 正誤 ☐☐	右の標識のある道路では，原動機付自転車は最も左側の車両通行帯を通行することはできない。
問34 正誤 ☐☐	右左折の合図をする時期は，右左折しようとする地点の30メートル手前に達したときである。（環状交差点を除く）
問35 正誤 ☐☐	原動機付自転車も，自賠責保険または責任共済に加入しなければならない。
問36 正誤 ☐☐	原動機付自転車は，車両通行帯の有無にかかわらず，トンネル内では走行中の自動車を追越ししてはならない。
問37 正誤 ☐☐	横の信号が赤になると同時に前方の信号が青に変わるので，前方の信号をよく見て速やかに発進しなければならない。
問38 正誤 ☐☐	歩行者用道路では，沿道に車庫を持つ車などで特に通行を認められた車だけが通行できる。
問39 正誤 ☐☐	交通渋滞のときなど，前の車に乗っている人が急にドアを開けたり，歩行者が車の間から飛び出すことがあるので注意が必要である。
問40 正誤 ☐☐	右の標示は「自転車専用道路」であることを表している。
問41 正誤 ☐☐	制動距離は，車の速度に二乗に比例して長くなる。
問42 正誤 ☐☐	仲間の車と行き違う場合や車の到着を知らせる場合は，警音器を鳴らしてもよい。
問43 正誤 ☐☐	追越しが禁止されている場所でも，原動機付自転車であれば追越ししてもよい。

158　第3回模擬試験　問題

Q29. T F ☐☐	In no-parking areas, you must not stop your vehicle, even for a short time, to pick up or drop off passengers.
Q30. T F ☐☐	When a tire punctures, immediately applying the brakes to stop is necessary.
Q31. T F ☐☐	Applying the brakes strongly allows the vehicle to stop within a short distance.
Q32. T F ☐☐	When attempting to overtake, always check for signs or markings to ensure that the area is not designated as a no-overtaking zone.
Q33. T F ☐☐	On roads where there is a indicated sign, motorized bicycles are not allowed to use the leftmost vehicle lane for traffic.
Q34. T F ☐☐	The appropriate timing to signal for a right or left turn is when reaching a point 30 meters before the intended turn (excluding roundabouts).
Q35. T F ☐☐	Motorized bicycles must also be covered by compulsory insurance or liability mutual aid.
Q36. T F ☐☐	Motorized bicycles, regardless of the presence of vehicle lanes, must not overtake moving vehicles in tunnels.
Q37. T F ☐☐	When the side signal turns red and simultaneously the front signal turns green, you must promptly check the front signal and proceed without delay.
Q38. T F ☐☐	In pedestrian zones, only vehicles specially permitted for passage, such as those with garages along the roadside, are allowed to pass.
Q39. T F ☐☐	During traffic jams, there is a possibility that passengers in the preceding vehicle may suddenly open their doors, or pedestrians may emerge from between cars, so caution is necessary.
Q40. T F ☐☐	The right sign indicates that it is a "bicycle lane."
Q41. T F ☐☐	The braking distance increases proportionally with the square of the vehicle's speed.
Q42. T F ☐☐	When passing friend's vehicle or to announce the arrival of a vehicle, it is permissible to use the horn.
Q43. T F ☐☐	Even in areas where passing is prohibited, it is permitted to overtake bicycles.

問44 正誤 □□	交差点の信号が黄色に変わったとき、停止位置に近づきすぎていて急ブレーキをかけなければ停止できないような場合は、そのまま進める。
問45 正誤 □□	タイヤがすり減っていると、摩擦抵抗が小さくなり、停止距離が長くなる。
問46 正誤 □□	原動機付自転車は、標識などによって路線バスの専用通行帯が指定されている道路を通行することができる。

問47　前方の工事現場の側方を対向車が直進してきます。この場合、どのようなことに注意して運転しますか？

(1) 正誤 □□　急にとまると、後ろの車に追突されるかもしれないので、ブレーキを数回に分けてかけ、停止の合図をする。

(2) 正誤 □□　工事現場から急に人が飛び出してくるかもしれないので、注意しながら走行する。

(3) 正誤 □□　対向車が来ているので、工事現場の手前で一時停止し、対向車が通過してから発進する。

問48　時速30キロメートルで進行しています。交差点を左折するときは、どのようなことに注意して運転しますか？

(1) 正誤 □□　前車はガソリンスタンドに入るかどうか分からないので、十分に車間距離を保ち、その動きに注意して進行する。

(2) 正誤 □□　前車はガソリンスタンドに入ると思われるので、右の車線に移り、前車を追い越して、左折する。

(3) 正誤 □□　前車も交差点を左折すると思うので、前車に接近して左折する。

Q44. T F ☐☐	When the traffic light changes to yellow at an intersection and you are too close to the stopping position to stop safely without sudden braking, you may proceed through the intersection.
Q45. T F ☐☐	When the tires are worn, the friction resistance decreases, leading to longer stopping distances.
Q46. T F ☐☐	Motorized bicycles are allowed to travel on roads where bus lanes are designated for exclusive use by signs or markings.

Q47. Oncoming car is approaching straight from the side of the construction site ahead. In this case, what should you pay attention to while driving?

(1) T F ☐☐ If you suddenly stop, there's a risk of being rear-ended by the car behind, so brake gradually in stages and signal your stop.

(2) T F ☐☐ Be cautious while driving as someone might suddenly emerge from the construction site.

(3) T F ☐☐ Stop momentarily before the construction site as oncoming traffic approaches, then proceed after the oncoming vehicles have passed.

Q48. You are proceeding at a speed of 30 kilometers per hour. what precautions do you take while driving?

(1) T F ☐☐ Maintain a sufficient distance from the vehicle in front, as it may be entering the gas station, and proceed with caution, keeping an eye on its movements

(2) T F ☐☐ Move to the right lane as the preceding vehicle appears to be entering the gas station, overtake it, and then make the left turn.

(3) T F ☐☐ Approach the intersection closely behind the preceding vehicle as it also seems to be making a left turn.

第3回 解答と解説	間違えたら赤シートを当てて，覚えておきたいポイントを再チェックしよう！

◆・・・ひっかけ問題　　★・・・重要な問題

問1 正	ミニカーとは総排気量 50cc 以下または定格出力 600 ワット 以下の原動機を有する小型の普通自動車のことをいう。
問2 ★誤	交通巡視員の手信号と信号機の信号とが違っている場合は，交通巡視員に従う。
問3 正	マナーモードやドライブモード，電源を切るなどして，呼び出し音が鳴らないようにする。
問4 誤	数値で表すのではなく，徐行とは車がすぐに停止できる速度で進行すること。
問5 ◆誤	交通規制を守るだけではなく，周囲の人の立場も考えて通行する。
問6 正	警察官の身体の正面に対面する方向は赤色，平行する方向は黄色の灯火信号と同じ意味である。
問7 誤	衝撃力とは速度の二乗に比例する。なので，4分の1である。
問8 誤	原動機付自転車がリヤカーなどけん引する場合は，最高速度は時速 25 キロメートルである。
問9 ★誤	前方に学校，幼稚園，保育所などがあることを意味する標識である。
問10 誤	トンネルに出入りするときは，速度を落とすようにする。
問11 ★正	原動機付自転車の積み荷の高さの制限は，地上から2メートル以下。
問12 正	原動機付自転車では，二輪と標示してある停止線の手前で停止する。
問13 正	強制保険が満了する年，月を示している。

The Third Practice Test Answers and Explanations	If you make a mistake, use a red sheet to cover it and re-check the points you want to remember!

◆ · · · Tricky question   ★ · · · Important question

Q1. T	A minicar refers to a small regular automobile with a total displacement of **50 cc** or less or a rated output of **600 watts** or less for the engine.
Q2. ★ F	If the hand signals of a traffic warden differ from the signals of a traffic light, you should follow the directions of the traffic warden.
Q3. T	Put your device on silent mode, drive mode, or turn it off to avoid hearing any incoming calls or notifications.
Q4. F	"Driving at Reduced Speed" refers to the speed at which a car moves slowly and can quickly come to **a stop**, rather than being expressed as a numerical value.
Q5. ◆ F	Not only should traffic regulations be followed, but also consideration should be given to the perspectives of those around when traveling.
Q6. T	Facing the front of a police officer's body indicates the same meaning as **a red signal light**, while facing parallel to their body signifies the same as **a yellow signal light**.
Q7. F	The force of impact is **proportional** to the square of the velocity. Therefore, it is one-fourth.
Q8. F	When a motorized bicycle is towing a trailer, such as a rear cart, the maximum speed is **25 kilometers per hour**.
Q9. ★ F	The sign indicates the presence of a school, kindergarten, or nursery ahead.
Q10. F	When entering or exiting a tunnel, it is advisable to reduce **speed**.
Q11. ★ T	The height limit for the load of a motorized bicycle is **2 meters** or less from the ground.
Q12. T	In the case of a motorized bicycle, you should stop **at the stop line** marked with motorcycle
Q13. T	It indicates the year and month when compulsory insurance expires.

The Third Practice Test Answers and Explanations   163

問14 正	できる限り道路の**左端**に寄り，環状交差点の側端に沿って**徐行**しながら通行する。
問15 誤	ブレーキ装置に水が入ると，一時的にブレーキの効きが悪くなる場合がある。
問16 誤	**駐車**が禁止されていて，**停車**は禁止されていない。
問17 ★正	霧の中は危険なので，慎重に運転する。
問18 誤	黄色の線の車両通行帯がある道路を通行していても，緊急自動車が近づいてきたら，道路の**左側**に寄って進路を譲らなければならない。
問19 正	盗難にも注意していた方が良い。
問20 誤	このような場合は，追越ししてはいけない。
問21 ★正	小型特殊自動車や原動機付自転車に加えて，**軽車両**も通行することができる。
問22 正	**指定方向外進行禁止**の標識で，右折はできない。
問23 ★正	夜間の運転時に反射性のついたものを着用すると，**危険防止**になる。
問24 正	危険を避けるためにやむを得ないときは，割り込んでもよい。
問25 誤	追越しの場合でも，最高速度を**超えて**はいけない。
問26 正	ブレーキを数回に分けて踏むと，制動灯の点滅により後続車の**追突防止**になる。
問27 ★誤	この標示は**駐車禁止**の場所を示している。

Q14. T	Keep as close to the **left side** of the road as possible and drive **at reduced speed** along the edge of the roundabout.
Q15. F	Water entering the brake system can temporarily reduce the effectiveness of the brakes.
Q16. F	**Parking** is prohibited, but **stopping** is not prohibited.
Q17. ★ T	Driving in fog is dangerous, so proceed with caution.
Q18. F	Even if there is a yellow-lined lane, when an emergency vehicle approaches, you must yield by moving to the **left side** of the road.
Q19. T	It's also advisable to be cautious of theft.
Q20. F	In such a case, overtaking is prohibited.
Q21. ★ T	Special light equipments, motorized bicycles, and **light road vehicles** are also allowed to pass.
Q22. T	The sign indicating " **Proceed Only in the Designated Direction**" prohibits right turns.
Q23. ★ T	Wearing clothing that reflects light during nighttime driving helps **improve safety**.
Q24. T	In unavoidable situations where safety is at risk, it may be necessary to cut in.
Q25. F	Even when overtaking, you must not **exceed** the speed limit.
Q26. T	Applying the brakes in several steps activates the brake lights, which helps **prevent rear-end collisions** from following vehicles.
Q27. ★ F	This sign indicates **a no-parking** zone.

The Third Practice Test Answers and Explanations   165

問28 誤	警察官の手信号などに従い，信号機の信号には従わない。
問29 ◆正	人待ちや，荷物待ちのたとえわずかな時間でも駐車にあたるので，**駐車禁止**である。
問30 誤	エンジンブレーキをかけながら，徐々に**速度**を落とす。
問31 ◆誤	ブレーキを強くかけると車輪の回転が止まり，スリップしてしまうおそれがあり危険なので，**停止距離**も短くなるとは限らない。
問32 ★正	標識や標示を確かめること。
問33 誤	大型貨物自動車，中型貨物自動車（車両総重量８トン未満を除く），大型特殊自動車に対しての通行帯の標識だが，原動機付自転車も**通行**できる。
問34 ★正	右左折する合図は，右左折する地点の 30 メートル手前に達したときに出す。
問35 正	自賠責保険，責任共済に加入しなければならない。
問36 ◆誤	車両通行帯があるときは，**追い越し**できる。
問37 ◆誤	交差点には一時的に全部赤になるところもあるので，必ず前方の信号を見るようにする。
問38 正	認められた車以外は，歩行者用道路は通行できない。
問39 正	交通渋滞のときでも，前の車や歩行者に注意をする。
問40 ★誤	「**自転車横断帯**」の標示であり，自転車が道路を横断するための場所である。
問41 正	制動距離とは，速度の**二乗**に比例して長くなる。

Q28. F	Follow the hand signals of the police officer, ignoring the signals from the traffic lights.
Q29. ◆ T	Even a short period of time, such as waiting for someone or waiting for a package, is considered parking, so it's considered **parking prohibited**.
Q30. F	Reduce **speed** gradually while applying the engine brake.
Q31. ◆ F	Braking too hard may cause the wheels to lock up and skid, which can be dangerous, and it doesn't necessarily shorten **the stopping distance**.
Q32. ★ T	Check the signs and markings.
Q33. F	The sign indicates a lane for large freight vehicles, medium-sized freight vehicles (excluding those with a total weight of less than 8 tons), and large special-purpose vehicles, but **it also allows** the passage of motorized bicycles.
Q34. ★ T	The signal for turning right or left should be given when reaching a point 30 meters before the intended turn.
Q35. T	You must get compulsory automobile liability insurance and liability mutual aid.
Q36. ◆ F	When there is a vehicle lane, **overtaking** is allowed.
Q37. ◆ F	Some intersections may have all-red signals temporarily, so always make sure to check the signal ahead.
Q38. T	Only authorized vehicles are allowed to pass on pedestrian roads.
Q39. T	Even in traffic jams, it's important to pay attention to the vehicles and pedestrians ahead.
Q40. ★ F	It indicates a "**Bicycle Crossing**" zone, where bicycles cross the road.
Q41. T	The braking distance increases proportionally **with the square** of the velocity

The Third Practice Test Answers and Explanations 167

問42 誤	警笛の乱用にあたるため，鳴らしてはいけない。
問43 誤	原動機付自転車であっても，追い越しはできない。
問44 正	安全に停止できない場合は，そのまま**進行**してよい。
問45 正	タイヤの状態がよい場合に比べて，タイヤがすり減ると**停止距離**が長くなる。
問46 正	身体の不自由な人が歩いているときは，**一時停止**や**徐行**で安全に通れるようにする。

問47　**バックミラー**に映る後続車と**対向車**に要注目！！
対向車がいるときは，無理やり通過すると激突するおそれがあるので，注意。減速するときも，後続車の追突に気をつけよう。

(1)　正　後続車の追突防止のため，ブレーキを数回に分けてかけて，**停止**します。
(2)　正　飛び出してくる人に十分に気をつけて**走行**する。
(3)　正　手前で一時停止して，**対向車**を先に行かせる。

問48　前車の**運転行動**に要注意！！
前車が左側のガソリンスタンドに入るのか，その先の交差点を左折するかをよく確認して運転しよう。

(1)　正　前車の**動き**によく注意して進行します。
(2)　誤　交差点の直前では**追越し**してはいけない。
(3)　誤　前車はガソリンスタンドに入るために，いきなり速度を**落**とすおそれがある。

Q42. F	You should not sound the horn excessively, as it constitutes misuse.
Q43. F	Even for motorized bicycles, passing is not allowed.
Q44. T	If you cannot safely stop, you may continue **forward**.
Q45. T	When tires are worn compared to when they are in good condition, **the stopping distance** becomes longer.
Q46. T	When persons with the physical disabilities are walking, **reduce speed** or **stop** to ensure safe passage.

Q47.   Pay attention to both the vehicles behind you and **oncoming traffic** by **the rearview mirror**. When there is oncoming traffic, avoid attempting to pass forcefully not to clash with each other. Also, be cautious of rear-end collisions when slowing down.

(1)   T   To prevent rear-end collisions, apply the brakes gradually in stages to come to a **stop**.
(2)   T   **Be vigilant** and watch out for pedestrians who may suddenly step into the road.
(3)   T   Come to a stop beforehand and allow **oncoming traffic** to proceed first.

Q48.   Be vigilant of **the driving behavior** of the vehicle in front!
Pay close attention to whether the vehicle ahead is turning into the left-side gas station or making a left turn at the upcoming intersection before proceeding with your own driving.

(1)   T   Proceed with caution, paying close attention to **the movements** of the vehicle in front
(2)   F   Avoid **overtaking** just before the intersection.
(3)   F   Cars may suddenly **reduce** speed to enter the gas station.

The Third Practice Test Answers and Explanations   169

第4回 模擬試験	問1〜問46までは各1点，問47，48は全て正解して各2点。制限時間30分，50点中45点以上で合格

●次の問題で正しいものは「正」，誤りのものは「誤」の枠をぬりつぶして答えなさい。

問1 正誤 □□	原動機付自転車の積載装置に積むことのできる荷物の長さは，荷台の長さに0.3メートル以下を加えた長さである。
問2 正誤 □□	ぬかるみや砂利道などを通過するときは，速度を上げて一気に通過するとよい。
問3 正誤 □□	通行に支障のある高齢者のそばを通るときは，一時停止か徐行しなければならない。
問4 正誤 □□	原付免許では，原動機付自転車と小型特殊自動車を運転することができる。
問5 正誤 □□	横断歩道のない交差点の手前で歩行者が横断中だったが，警音器を鳴らしたら横断をやめたので，そのまま進行した。
問6 正誤 □□	交差点を右折するときは，自分の車が先に交差点に入っていても，反対方向からの直進車や左折車の進行を妨げてはならない。
問7 正誤 □□	夜間に，自車のライトと対向車のライトで道路の中央付近の歩行者や自転車が見えなくなることを「蒸発現象」という。
問8 正誤 □□	トンネルに入るときは減速するが，トンネルから出るときは減速する必要はない。
問9 正誤 □□	原動機付自転車が右折しようとするとき，右の図の矢印のような進路をとる。 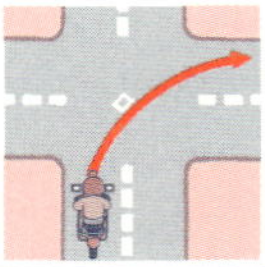
問10 正誤 □□	ひとり歩きしているこどものそばを通るときは，1メートルくらいの間隔をあければ特に徐行などしなくてよい。
問11 正誤 □□	同乗者用座席がない普通自動二輪車や，原動機付自転車では，二人乗りはしてはいけない。
問12 正誤 □□	右の図の標識がある道路では，原動機付自転車の最高速度は時速40キロメートルである。 
問13 正誤 □□	交通整理をしている警察官が，正面を向いて腕を水平に上げていたので，その交差点を左折した。

170　第4回模擬試験　問題

Fourth Practice Test	Questions 1 to 46 are worth 1point each, while questions 47 and 48 are worth 2 points each if all answers are correct. The time limit is 30 minutes. To pass, you need 45 points out of 50.	

● Fill in the box marked 'T' for the correct answers and 'F' for the incorrect ones in the following questions.

Q1. T F ☐☐	The length of the luggage that can be loaded on the loading device of a motorized bicycle is the length of the loading platform plus 0. 3 meters or less.
Q2. T F ☐☐	When passing through muddy or gravel roads, it is advisable to increase speed and pass through quickly.
Q3. T F ☐☐	When passing near elderly people who may obstruct traffic, you must either come to a temporary stop or drive at reduced speed.
Q4. T F ☐☐	With a motorized bicycle license, you are allowed to drive both motorized bicycles and special light equipment.
Q5. T F ☐☐	At an intersection without a pedestrian crossing, pedestrians were crossing just before the intersection, but they stopped crossing when I sounded the horn, so I continued driving.
Q6. T F ☐☐	When turning right at an intersection, even if your vehicle has already entered the intersection, you must not obstruct the passage of oncoming straight or left-turning vehicles.
Q7. T F ☐☐	- At night, when pedestrians or cyclists in the middle of the road become invisible due to the headlights of your vehicle and oncoming vehicles, it is referred to as the "Glare".
Q8. T F ☐☐	When entering a tunnel, it is necessary to reduce speed, but when exiting a tunnel, there is no need to reduce speed.
Q9. T F ☐☐	When a motorized bicycle intends to turn right, it should take a path similar to the arrow in this diagram.
Q10. T F ☐☐	When passing by a child walking alone, it is not necessary to reduce speed if you maintain a distance of about one meter.
Q11. T F ☐☐	On motorcycles without passenger seats or on motorized bicycles, carrying a passenger is prohibited.
Q12. T F ☐☐	On roads with the sign shown in the right diagram, the maximum speed limit for motorized bicycles is 40kilometers per hour.
Q13. T F ☐☐	A police officer directing traffic had their arm raised horizontally facing forward, so I turned left at that intersection.

Fourth Practice Test   171

問14 正誤 ☐☐	ブレーキは，ハンドルを切らない状態で車体を垂直に保ちながら，前後輪ブレーキを同時にかけるのがよい。
問15 正誤 ☐☐	標識や標示で最高速度が指定されていないところでは，法令で定められた最高速度を超えて原動機付自転車を運転してはならない。
問16 正誤 ☐☐	原動機付自転車は，車道が混雑しているときは路側帯を通行することができる。
問17 正誤 ☐☐	坂道での行き違いは，上り坂での発進が難しいので，下りの車が上りの車に道を譲る。
問18 正誤 ☐☐	事故を起こさない自信があれば，走行中に携帯電話を使用してもよい。
問19 正誤 ☐☐	第一種免許は，大型免許，中型免許，準中型免許，普通免許，大型二輪免許，普通二輪免許，原付免許の7種類である。
問20 正誤 ☐☐	進路変更の合図と右左折の合図の時期は同じである。
問21 正誤 ☐☐	標示とは，ペイントや道路びょうなどで路面に示された線や記号や文字のことで，規制標示と指示標示の2種類がある。
問22 正誤 ☐☐	右の図のような交通整理が行われていない道幅が同じような交差点では，Aの原動機付自転車はBの普通自動車に進路を譲らなければならない。 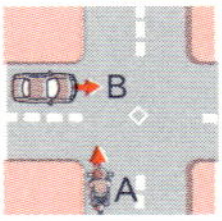
問23 正誤 ☐☐	前の自動車がその前の原動機付自転車を追い越そうとしているとき，その自動車を追い越そうとするのは二重追越しとなる。
問24 正誤 ☐☐	踏切直前で発進したときは，速やかにギアチェンジして，高速ギアで通過するようにしたほうがよい。
問25 正誤 ☐☐	火災報知機から1メートル以内の場所は，停車はできるが駐車はできない。
問26 正誤 ☐☐	小型特殊自動車の積載物の重量制限は，500キログラムである。
問27 正誤 ☐☐	右の標識のある場所では，午前8時から午後8時まで駐車してはならない。 
問28 正誤 ☐☐	夜間，街路灯などで明るい繁華街を走るときは，前照灯をつける必要はない。

Q14. T F ☐☐	When braking, it's best to keep the vehicle upright without steering and apply both front and rear brakes simultaneously.
Q15. T F ☐☐	In places where the maximum speed limit is not specified by signs or markings, motorized bicycles must not exceed the maximum speed limit prescribed by law.
Q16. T F ☐☐	Motorized bicycles are allowed to use the roadside when the roadway is congested.
Q17. T F ☐☐	When passing on a slope, vehicles going downhill should yield to vehicles going uphill because starting on an incline can be difficult.
Q18. T F ☐☐	If you are confident in your ability to avoid accidents, you may use a mobile phone while driving.
Q19. T F ☐☐	The Class 1 driver's license includes seven types: Large motor vehicle, Medium-sized motor vehicle, Semi-medium-sized motor vehicle, Standard motor vehicle, Large Motorcycle, Standard Motorcycle, and Motorized bicycle licenses.
Q20. T F ☐☐	The timing for signaling lane changes is the same as signaling for turning right or left.
Q21. T F ☐☐	Markings refer to lines, symbols, and characters painted or marked on the road surface, including regulatory and advisory markings.
Q22. T F ☐☐	At intersections with equally wide roadways and no traffic control like the diagram on the right, Moped A must yield to Car B. 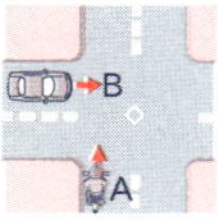
Q23. T F ☐☐	Overtaking a vehicle that is attempting to overtake the motorized bicycle in front of it constitutes double overtaking.
Q24. T F ☐☐	When starting just before a railroad crossing, it's advisable to quickly change gears and pass through in a high gear.
Q25. T F ☐☐	Parking is not allowed within 1meter of a fire alarm.
Q26. T F ☐☐	The weight limit for cargo on a special light equipment is 500 kilograms.
Q27. T F ☐☐	Parking is prohibited from 8a. m. to 8 p. m. in areas with the sign shown on the right. 
Q28. T F ☐☐	When driving through well-lit urban areas at night, it is not necessary to turn on the headlights.

Fourth Practice Test 173

問29 正誤 ☐☐	対向車と行き違うときは，安全な間隔を保たなければならない。
問30 正誤 ☐☐	交通事故を起こしても，任意保険に加入していれば，民事上の責任はすべて保険会社が負うこととなる。
問31 正誤 ☐☐	駐停車禁止場所では，原則として車の駐車や停車が禁止されているが，危険防止のためやむを得ず一時停止するようなときは，停止できる。
問32 正誤 ☐☐	信号機のある踏切で，青色の灯火を示しているときは，一時停止しないで通過することができる。
問33 正誤 ☐☐	右の信号の赤色の灯火の点滅が表示されているとき車は，ほかの交通に注意して進むことができる。　〇〇●
問34 正誤 ☐☐	発進するときは，前後の交通の安全を確かめて，右側の方向指示器を作動するか手で合図をしなければならない。
問35 正誤 ☐☐	前方の信号が黄色のときは，ほかの交通に注意しながら進行することができる。
問36 正誤 ☐☐	酒を飲んでいるのを知りながら原動機付自転車を運転して配達することを依頼したときは，依頼した人も罰せられることがある。
問37 正誤 ☐☐	下り坂では加速がつくため，高速ギアを用いてエンジンブレーキを活用する。
問38 正誤 ☐☐	進行方向別通行区分に従って通行しているときに緊急自動車が近づいてきた場合は，その通行区分が終わってから進路を譲ればよい。
問39 正誤 ☐☐	初心者マークや高齢運転者マークをつけている自動車を，追い越すことはできない。
問40 正誤 ☐☐	右の標識のある道路では，自動車は通行できないが原動機付自転車は通行できる。
問41 正誤 ☐☐	追越しが禁止されていない，左側部分の幅が６メートル未満の見通しのよい道路で，ほかの車を追い越そうとするとき，道路の中央から右側部分に最小限はみ出して通行することができる。
問42 正誤 ☐☐	ごく少量の酒を飲んだが，酔っていないので慎重に運転した。
問43 正誤 ☐☐	右の標識によって路線バス専用通行帯が指定されている道路でも，原動機付自転車は通行することができる。

174　第４回模擬試験　問題

Q29. T F ☐☐	When passing each other with oncoming vehicles, a safe distance must be maintained.
Q30. T F ☐☐	Even if a traffic accident occurs, if you have voluntary insurance, the insurance company will bear all civil liability.
Q31. T F ☐☐	In no parking zones, parking or stopping of vehicles is generally prohibited, but temporary stops may be allowed if necessary for safety reasons.
Q32. T F ☐☐	At a railroad crossing with traffic lights, you can pass without stopping when the green light is displayed.
Q33. T F ☐☐	When the red light is flashing at the signal on the right, vehicles may proceed with caution while observing other traffic.
Q34. T F ☐☐	When starting, you must ensure the safety of surrounding traffic and activate the right turn signal or signal with your hand.
Q35. T F ☐☐	When the signal ahead is yellow, you can proceed while paying attention to other traffic.
Q36. T F ☐☐	If someone requests you to operate a motorized bicycle while knowing you are drinking alcohol, the requester may also be punished.
Q37. T F ☐☐	On downhill slopes, using the engine brake with a high gear is effective due to the acceleration.
Q38. T F ☐☐	When following the directional lanes and an emergency vehicle approach, you should yield after the lane ends.
Q39. T F ☐☐	It is not allowed to overtake vehicles displaying beginner or elderly driver marks.
Q40. T F ☐☐	On roads with the sign on the right, cars are not allowed, but mopeds are permitted.
Q41. T F ☐☐	When overtaking on a clear road with a width of less than 6 meters and no prohibition on overtaking, vehicles may slightly protrude into the right lane from the center of the road.
Q42. T F ☐☐	Although I drank a very small amount of alcohol, I was not intoxicated, so I drove carefully.
Q43. T F ☐☐	Even on roads designated as bus lanes by the sign on the right, mopeds are allowed to travel.

Fourth Practice Test   175

問44 正誤 □□	昼間でも，トンネルの中や，50メートル先が見えないような場所を通行するときは，前照灯などをつけなければならない。
問45 正誤 □□	上り坂の頂上付近は，徐行の標識がなければ徐行する必要がない。
問46 正誤 □□	交差点（環状交差点を除く）で右折しようとして自分の車が先に交差点に入ったときは，その交差点の対向車線を直進してくる車より先に進行できる。

問47 交差点の手前で停止しました。渋滞している交差点を直進するときは，どのようなことに注意して運転しますか？

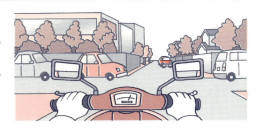

(1) 正誤 □□ 進行方向の渋滞している車の間はあいているので，交差点に入る前に左右を確認したらすばやく通過する。

(2) 正誤 □□ 渋滞している車が動き出すおそれがあるので，交差点に入るときは，渋滞している先のほうを確認してから発進する。

(3) 正誤 □□ 渋滞している車の向こう側から二輪車が走行してくるかもしれないので，その手前で止まって左側を確かめながら通過する。

問48 時速30キロメートルで進行しています。この場合，どのようなことに注意して運転しますか？

(1) 正誤 □□ 歩行者がバスのすぐ前を横断するかもしれないので，いつでも止まれるような速度に落として，バスの側片を通過する。

(2) 正誤 □□ バスを降りた人が，バスの後ろを横断するかもしれないので，警音器を鳴らし，いつでもハンドルを右に切れるよう注意して進行する。

(3) 正誤 □□ 対向車が来るかどうかバスのかげでよく分からないので，前方の安全をよく確かめてから，中央線を越えて進行する。

Q44. T F ☐☐	Even during the daytime, when passing through tunnels or areas where visibility is less than 50 meters, you must turn on your headlights.
Q45. T F ☐☐	Near the top of an uphill slope, there is no need to reduce speed unless there are signs indicating so.
Q46. T F ☐☐	When attempting to make a right turn at an intersection (excluding roundabouts) and your vehicle enters the intersection first, you may proceed before vehicles from the opposing lane intending to proceed straight.

Q47. You stopped before the intersection. What should you be careful of when driving straight through a congested intersection?

(1) T F ☐☐ There is a gap between the cars in traffic in the direction of travel, so after checking both left and right before entering the intersection, You will quickly pass through.

(2) T F ☐☐ Since there's a possibility that the congested cars might start moving, you will check ahead in the direction of the congestion before proceeding into the intersection.

(3) T F ☐☐ There might be motorcycles coming from the other side of the congested cars, so you will stop before that and pass through while checking the left side.

Q48. When traveling at a speed of 30 kilometers per hour, what should you pay attention to while driving in this situation?

(1) T F ☐☐ Pedestrians may cross right in front of the bus, so reduce your speed to a level where you can stop at any time and pass the side of the bus.

(2) T F ☐☐ People getting off the bus might cross behind it, so be cautious, sound the horn, and be ready to turn right at any moment.

(3) T F ☐☐ It's hard to see if oncoming traffic is approaching due to the bus blocking the view, so you carefully check the safety ahead before crossing the center line.

Fourth Practice Test 177

第4回 解答と解説	間違えたら赤シートを当てて，覚えておきたいポイントを再チェックしよう！

◆・・・ひっかけ問題　　★・・・重要な問題

問1 正	荷台の長さに**0.3メートル**以下を加えた長さである。
問2 ◆誤	ぬかるみや砂利道では，低速ギアを使い，**減速**して通行するのがよい。
問3 正	高齢者が安全に通行できるように，**一時停止**か**徐行**する。
問4 誤	原付免許では，原動機付自転車のみ運転できる。
問5 誤	横断歩道のない交差点などを歩行者が横断しているときは，その**通行**を妨げない。
問6 正	先に交差点に入っていても，**直進車**や**左折車**の進行を妨げてはいけない。
問7 ★正	歩行者などが見えなくなることを**蒸発現象**といい，十分に注意する。
問8 ◆誤	トンネルなどで明るさが急に変わると，一時的に視力が低下して見えにくくなるので，出るときも**速度**を落とす。
問9 誤	あらかじめ道路の**中央**に寄り，右折しなければならない。
問10 ★誤	子どもがひとり歩きしている場合は，安全に通れるように**一時停止**か**徐行**する。
問11 正	乗車定員は，1人のみ。
問12 ◆誤	この標識があっても，原動機付自転車は時速**30キロメートル**を超えて運転してはいけない。
問13 誤	警察官の手信号は，信号機の赤色の灯火を表しているので，直進や右左折はできない。

| | The Fourth Practice Test Answers and Explanations | If you make a mistake, use a red sheet to cover it and re-check the points you want to remember! |

◆ · · · Tricky question   ★ · · · Important question

Q1. T	It is the length obtained by adding **0. 3 meters** or less to the length of the loading platform.
Q2. ◆ F	In muddy or gravel roads, it is advisable to use low gear and **decelerate** for passage.
Q3. T	To ensure safe passage for the elderly, come to **a temporary stop** or drive **at reduced speed**.
Q4. F	With a motorized bicycle license, you can operate only a motorized bicycle.
Q5. F	When pedestrians are crossing at intersections without crosswalks, do not obstruct **their passage**.
Q6. T	Even if one's vehicle has entered the intersection first, do not obstruct the progress of **vehicles traveling straight** or **turning left**.
Q7. ★ T	The phenomenon where pedestrians and other objects become invisible is called "**Glare**". Be cautious.
Q8. ◆ F	In tunnels where brightness changes abruptly, temporary visual impairment occurs, making it difficult to see, so reduce **speed** when exiting.
Q9. F	Prior to turning right, one must first move closer to **the center** of the road.
Q10. ★ F	When a child is walking alone, come to **a temporary stop** or drive **at reduced speed** to ensure safe passage.
Q11. T	The maximum occupancy is one person only.
Q12. ◆ F	Even with this sign, motorized bicycles must not exceed **30 kilometers per hour**.
Q13. F	Police officers' hand signals represent the red lights of traffic signals, so one cannot proceed straight or turn right or left.

The Fourth Practice Test Answers and Explanations   **179**

問14 ◆正	車体を垂直に保ち，前後輪ブレーキを同時にかける。
問15 ★正	最高速度が指定されていないところでは，法令で定められている最高速度を**超えて**はいけない。
問16 ★誤	原則として，**路側帯**は歩行者の通行するところである。
問17 正	原則として，下りの車が上りの車に道を譲る。
問18 誤	走行中は危険なので，携帯電話を使用してはいけない。反則金が科せられる。
問19 誤	大型特殊免許，小型特殊免許，けん引免許を加えた**10**種類である。
問20 ★誤	進路変更の合図は約**3秒**前，右左折の合図は右左折地点の**30メートル**手前である。（環状交差点を除く）
問21 正	ペイントや道路びょうなどで路面に示された線や記号，文字のことを**道路標示**という。
問22 正	左から進行してくる車の進行を妨げてはならないので，原動機付自転車は普通自動車に進路を譲る。
問23 ★誤	自動車ではない原動機付自転車が前の車を追い越そうとしているので，前の車を追い越しても**二重追越し**にはならない。
問24 誤	踏切内でエンストしないために，発進したときの低速ギアのまま一気に通過する。
問25 正	火災報知機から**1メートル**以内の場所は，**停車**できて**駐車**はできない。
問26 ★正	**500キログラム**まで小型特殊自動車には荷物を積める。
問27 ★正	標識に記載されている午前8時から午後8時まで**駐車禁止**。

Q14. ◆ T	Ensure the vehicle body remains perpendicular and apply both front and rear brakes simultaneously.
Q15. ★ T	In areas where maximum speed limit is not specified, one must not **exceed** the maximum speed limit prescribed by law.
Q16. ★ F	As a rule, **side strips** are designated for pedestrian traffic.
Q17. T	As a general rule, vehicles traveling downhill should yield to those traveling uphill.
Q18. F	Using a mobile phone while driving is dangerous and punishable by a fine.
Q19. F	There are a total of 10 types of licenses, including the special heavy equipment license, small special light equipment license, and towing license.
Q20. ★ F	The signal for changing lanes should be given approximately **3 seconds** before, while the signal for turning right or left should be given **30 meters** before the turn (excluding roundabouts).
Q21. T	Lines, symbols, and characters painted or marked on the road surface are referred to as **road markings**.
Q22. T	Motorized bicycles must yield to regular automobiles to avoid obstructing the passage of vehicles approaching from the left.
Q23. ★ F	Even if a non-automobile motorized bicycle is attempting to pass the preceding vehicle, overtaking it does not constitute **double overtaking**.
Q24. F	To avoid stalling on a railway crossing, it is advisable to proceed through it swiftly while maintaining low gear from the moment of departure.
Q25. T	**Parking** is prohibited within **1meter** of a fire alarm, but **stopping** is allowed.
Q26. ★ T	Up to **500 kilograms** of cargo can be loaded onto a special light equipment.
Q27. ★ T	**Parking is prohibited** from 8:00a. m. to 8:00 p. m. as indicated on the sign.

The Fourth Practice Test Answers and Explanations

問28 誤	夜間は必ずライトをつける。
問29 正	**安全な間隔**を保つこと。
問30 誤	すべて運転者本人の責任になる。
問31 正	**危険防止**のための一時停止であれば，駐停車禁止場所でも停止できる。
問32 正	**一時停止**しないで，安全確認をして通過できる。
問33 誤	車は停止位置で**一時停止**して，安全確認しなければならない。
問34 正	発進するときは，**安全確認と合図**をして発進する。
問35 誤	停止位置で安全に停止できないとき以外は，停止位置を越えて進んではならない。
問36 正	酒を飲んでいるのを知りながら配達を依頼すると，依頼した人も罰せられる場合がある。
問37 誤	下り坂では**低速ギア**を使って，エンジンブレーキを活用する。
問38 誤	緊急自動車がきたときは，通行区分に**従う必要はない**ので進路を譲る。
問39 誤	追越しはできるが，幅寄せや割り込みは禁止されている。
問40 ★誤	「**車両通行止め**」の標識で歩行者以外の車両の通行を禁止している。
問41 正	このような道路では，道路の中央から右側に最小限はみ出して追越しできる。

Q28. F	Always turn on your lights at night.
Q29. T	Keep **a safe distance**.
Q30. F	The responsibility falls entirely on the driver.
Q31. T	It is permitted to stop temporarily in no-parking zones for **safety-related reasons**.
Q32. T	You may proceed without stopping after making safety checks.
Q33. F	Vehicles must come to **a temporary halt** at the stopping position and perform safety checks.
Q34. T	When starting, make **safety checks** and **signal** before proceeding.
Q35. F	Except when unable to safely stop at the stopping position, do not proceed beyond the stopping position.
Q36. T	If you request delivery while knowing the person is drinking, you may also be penalized.
Q37. F	Use **low gear** and engine braking on downhill slopes.
Q38. F	When an emergency vehicle approaches, you are **not required to follow** the lane of travel and should yield the right of way.
Q39. F	Overtaking is allowed, but crowding or cutting in is prohibited.
Q40. ★ F	The "**Road Closed**" sign prohibits the passage of vehicles other than pedestrians.
Q41. T	In such roads, you can overtake by protruding minimally to the right from the center of the road.

The Fourth Practice Test Answers and Explanations　183

問42 ★誤	少量でも**飲酒運転**になり，絶対に運転してはならない。
問43 正	原動機付自転車は通行することができる。
問44 正	昼間でもトンネルの中や，50メートル先が見えない場所では前照灯などをつけなければならない。
問45 誤	標識がなくても**徐行**しなければならない。
問46 誤	右折するときは，直進車や左折車の**進行**を妨げてはならない。

問47　**渋滞**の列の動きとその**かげ**に要注意！！
進路はあいていても，いつ車が**動き出す**か分かりません。また，車の**かげ**から二輪車が走行するかもしれないので注意しよう。

(1)　**誤**　車が急に**動き出す**かもしれない。
(2)　正　渋滞している先を確認して，**状況**に応じて判断する。
(3)　正　車のかげから二輪車が**走行してくる**かもしれません。

問48　**歩行者**と**対向車**の有無に要注意！！
バスを降りた人が道路を**横断**するかもしれないので注意して，バスのかげから出てくるかもしれない対向車にも気をつけよう。

(1)　正　歩行者がバスの前を**横断**するおそれがあります。
(2)　**誤**　歩行者の横断に注意して，**警音器**は鳴らさずに進行する。
(3)　正　**対向車**の接近に注意して進行する。

Q42. ★ F	Even a small amount constitutes **drunk driving** and must never be done.
Q43. T	Motorcycles with auxiliary engines are allowed to pass.
Q44. T	Even during the day, headlights must be turned on in tunnels or where visibility is less than 50 meters ahead.
Q45. F	Even without signs, you must **slow down** with caution.
Q46. F	When turning right, you must not obstruct **the progress** of oncoming or left-turning vehicles.

Q47.   Be cautious of the movement of the traffic **jam** queue and its **behind!**
Even if the path seems clear, you never know when a car will **start moving**. Also, be cautious as motorcycles may be traveling **behind** the car.

(1)   F   Cars may suddenly **start moving**.
(2)   T   Check ahead for traffic congestion and make decisions accordingly based **on the situation**.
(3)   T   Motorcycles may **be traveling** behind the cars.

Q48.   Be cautious of **pedestrians** and **oncoming vehicles!**
Watch out for pedestrians who may **cross** the road after getting off the bus, and also be mindful of oncoming vehicles that may emerge from behind the bus.

(1)   T   There is a risk of pedestrians **crossing** in front of the bus.
(2)   F   Be cautious of pedestrian crossings and proceed without sounding **the horn**.
(3)   T   Proceed with caution for approaching **oncoming vehicles**.

| 第5回 模擬試験 | 問1〜問46までは各1点，問47，48は全て正解して各2点。制限時間30分，50点中45点以上で合格 |

●次の問題で正しいものは「正」，誤りのものは「誤」の枠をぬりつぶして答えなさい。

問1 正誤 ☐☐	片側2車線の道路の交差点で原動機付自転車が右折するとき，標識による右折方法の指定がなければ，小回りの右折方法をとる。
問2 正誤 ☐☐	エンジンを止めた原動機付自転車を押して歩く場合でも，歩行者専用信号でなく，車両用信号に従って通行する。
問3 正誤 ☐☐	二輪車でブレーキをかける場合，路面が乾燥しているときは，後輪ブレーキをやや強めにかける。
問4 正誤 ☐☐	同一方向に進行しながら進路を右に変える場合，後続車がいなければ合図しなくてよい。
問5 正誤 ☐☐	原動機付自転車であっても，PS（c）マークやJISマークのヘルメットをかぶれば高速道路を通行することができる。
問6 正誤 ☐☐	乗降のため止まっている通学通園バスのそばを通るときは，1.5メートル以上の間隔をあけていれば，徐行しないで通過できる。
問7 正誤 ☐☐	原動機付自転車は，前方の信号が黄色や赤色であっても，青色の左折の矢印の信号の場合は，矢印の方向に進むことができる。
問8 正誤 ☐☐	普通車の仮免許では原動機付自転車を運転することはできない。
問9 正誤 ☐☐	右の標識は，優先道路を表している。
問10 正誤 ☐☐	原付免許を受けて1年間を初心運転者期間といい，この間に違反をして一定の基準に達した人は免許取り消しとなる。
問11 正誤 ☐☐	運転中は一点を注視しないで，前方のみを見渡す目配りをしたほうがよい。
問12 正誤 ☐☐	右の標識のあるところでは，原動機付自転車は徐行しなければならない。
問13 正誤 ☐☐	トンネルの中では，対向車に注意を与えるため，右側の方向指示器を作動させたまま走行した方がよい。

186　第5回模擬試験　問題

Fifth Practice Test	Questions 1 to 46 are worth 1point each, while questions 47 and 48 are worth 2 points each if all answers are correct. The time limit is 30 minutes. To pass, you need 45 points out of 50.

● Fill in the box marked 'T' for the correct answers and 'F' for the incorrect ones in the following questions.

Q1. T F ☐☐	When a motorized bicycle makes a right turn at an intersection on a two-lane road without designated turning instructions, it should make a tight right turn.
Q2. T F ☐☐	Even when walking with a motorized bicycle with the engine off, follow the vehicle signals, not the pedestrian signals.
Q3. T F ☐☐	When applying the brakes on a motorcycle, if the road surface is dry, apply the rear brake slightly more forcefully.
Q4. T F ☐☐	When changing lanes to the right while traveling in the same direction, signaling may not be necessary if there are no following vehicles.
Q5. T F ☐☐	Even on highways, wearing a helmet with a PS (c) mark or JIS mark allows motorized bicycles to travel.
Q6. T F ☐☐	When passing by a school bus stopped for boarding or alighting, you can pass without reducing speed if you maintain a distance of at least 1. 5 meters.
Q7. T F ☐☐	Even if the signal ahead is yellow or red, motorized bicycles can proceed in the direction of a green left-turn arrow signal.
Q8. T F ☐☐	A standard motor vehicle provisional license does not permit driving motorized bicycles.
Q9. T F ☐☐	The sign on the right indicates a priority road.
Q10. T F ☐☐	The first year after obtaining a motorized bicycle license is called the novice driver period. During this time, committing certain violations may result in license revocation.
Q11. T F ☐☐	While driving, it is better to scan the surroundings rather than fixating on one point.
Q12. T F ☐☐	Motorized bicycles must drive at reduced speed where the sign on the right is displayed.
Q13. T F ☐☐	In tunnels, it is advisable to keep the right turn signal on to alert oncoming traffic.

Fifth Practice Test 187

問14 正誤 ☐☐	対向車のライトがまぶしいときは，視線をやや左前方に移すようにする。
問15 正誤 ☐☐	ほかの車に追い越されるときに，追越しをするための十分な余地がないときは，できるだけ左に寄り進路を譲らなければならない。
問16 正誤 ☐☐	雨にぬれたアスファルトの路面では，車の制動距離は短くなるので，強くブレーキをかけるとよい。
問17 正誤 ☐☐	広い道路で右折をしようとするときは，左側車線から中央寄りの車線に一気に移動しなければならない。
問18 正誤 ☐☐	右折や左折をするときは，必ず徐行しなければならない。
問19 正誤 ☐☐	マフラーを改造していない原動機付自転車なら，著しく他人の迷惑になるような空ぶかしであっても禁止されていない。
問20 正誤 ☐☐	夜間，交通整理をしている警察官が頭上に灯火を上げているとき，その警察官の身体の正面に平行する交通については，青色の信号と同じ意味である。
問21 正誤 ☐☐	道路への出入口はもちろん，出入口から3メートル以内も駐車禁止である。
問22 正誤 ☐☐	右の標識のある道路では転回してはならない。	
問23 正誤 ☐☐	道路が混雑しているときに原動機付自転車で路側帯を通行した。
問24 正誤 ☐☐	バスの停留所の標示板（柱）から10メートル以内の場所では，停車はできるが駐車はできない。
問25 正誤 ☐☐	原動機付自転車の法定速度は30キロメートル毎時である。
問26 正誤 ☐☐	雨の降り始めの舗装道路や工事現場の鉄板などは，すべりやすいので注意したほうがよい。
問27 正誤 ☐☐	右の標識のある場所は危険物の貯蔵所などがあるので，注意して運転しなければならない。	

Q14. T F ☐☐	When the headlights of oncoming vehicles are dazzling, it's advisable to shift your gaze slightly to the left front.
Q15. T F ☐☐	When being overtaken by other vehicles and there is not enough space to safely pass, you must move to the left as much as possible and yield the lane.
Q16. T F ☐☐	On asphalt roads wet from rain, the braking distance of vehicles decreases, so it's advisable to apply the brakes firmly.
Q17. T F ☐☐	When making a right turn on a wide road, you must move from the left lane to the center lane all at once
Q18. T F ☐☐	When making a right or left turn, you must always drive at reduced speed.
Q19. T F ☐☐	Unless the muffler has modified, even significant noise emission that disturbs others is not prohibited.
Q20. T F ☐☐	At night, when a police officer raises a light overhead, it signifies the same as a green signal for traffic parallel to the officer's body front.
Q21. T F ☐☐	Parking is prohibited within 3meters of road entrances and exits, including the entrances and exits themselves.
Q22. T F ☐☐	Making U-turns is prohibited on roads with the sign on the right.
Q23. T F ☐☐	It's allowed to use the side strip with a motorized bicycle when roads are congested.
Q24. T F ☐☐	Within 10 meters of a bus stop signboard (pole), stopping is permitted but parking is not.
Q25. T F ☐☐	The statutory speed limit for motorized bicycles is 30 kilometers per hour.
Q26. T F ☐☐	It's advisable to be cautious on newly wet pavement roads due to rain or metal plates at construction sites as they can be slippery.
Q27. T F ☐☐	Areas with the sign on the right may have storage facilities for hazardous materials, so driving with caution is necessary.

Fifth Practice Test   189

問28 正誤 □□	カーブの手前では徐行しなければならない。
問29 正誤 □□	見通しのきく信号機がない踏切では，安全確認すれば一時停止する必要はない。
問30 正誤 □□	夜間は，視界が狭くなるので，できるだけ近くのものを見るようにする。
問31 正誤 □□	原動機付自転車は，交通が渋滞しているときでも，車の間をぬって走ることができるので便利である。
問32 正誤 □□	大地震が起き，車を置いて避難するときは，エンジンを止めてエンジンキーを確実に抜いておく。
問33 正誤 □□	右の標示のある交差点では，普通自転車は，この標示を超えて交差点に進入することは禁止されている。
問34 正誤 □□	停止位置に近づいたときに，信号が青色から黄色に変わったが，後続車があり急停止すると追突される危険を感じたので，停止せずに交差点を通り過ぎた。
問35 正誤 □□	水たまりを通過するときには，徐行するなどして歩行者などに泥水がかからないようにしなければならない。
問36 正誤 □□	坂の頂上付近は，駐車も停車も禁止されている。
問37 正誤 □□	原動機付自転車に積載することのできる荷物の重量限度は，30キログラムまでである。
問38 正誤 □□	車から離れるときでも，短時間であればエンジンを止めなくてもよい。
問39 正誤 □□	進路を変更すると，後ろからくる車が急ブレーキや急ハンドルで避けなければならないような場合は，進路を変えてはならない。
問40 正誤 □□	右の標示があるところでは，前方に優先道路がある。
問41 正誤 □□	前の車が交差点や踏切の手前で徐行しているときは，その前を横切ってはならないが，停止しているときは，その前を横切ってもよい。
問42 正誤 □□	道路を通行するときは，交通規則を守るほか，道路や交通の状況に応じて細かい注意をする必要がある。

Q28. T F ☐☐	Before approaching a curve, you must drive at reduced speed.
Q29. T F ☐☐	At a railroad crossing without a clear view of the signal, there is no need to stop if you can safely confirm there's no approaching train.
Q30. T F ☐☐	During the night, visibility decreases, so it's important to focus on nearby objects as much as possible.
Q31. T F ☐☐	Motorized bicycles can maneuver through traffic even when it's congested, making them convenient.
Q32. T F ☐☐	In the event of a major earthquake and evacuating the vehicle, it's crucial to turn off the engine and ensure the ignition key is removed.
Q33. T F ☐☐	At intersections with the sign on the right, standard bicycles are prohibited from entering the intersection beyond this sign.
Q34. T F ☐☐	When approaching a stop position and the signal changes from green to yellow but there are following vehicles, risking a rear-end collision by sudden braking, so I continue past the intersection without stopping.
Q35. T F ☐☐	When passing through puddles, it's necessary to drive at reduced speed to avoid splashing pedestrians with muddy water.
Q36. T F ☐☐	Parking or stopping is prohibited near the top of a hill.
Q37. T F ☐☐	The weight limit for cargo that can be loaded onto a motorized bicycle is up to 30 kilograms.
Q38. T F ☐☐	When stepping away from the vehicle for a short time, it's not necessary to turn off the engine.
Q39. T F ☐☐	When changing lanes, if a vehicle behind needs to brake or maneuver abruptly to avoid a collision, you must not change lanes.
Q40. T F ☐☐	Where there is the marking on the right, there is a priority road ahead.
Q41. T F ☐☐	When the preceding vehicle is driving at reduced speed near an intersection or railroad crossing, you must not attempt to overtake it. However, it's permissible to pass it when it's stationary.
Q42. T F ☐☐	While driving on the road, besides following traffic regulations, it's essential to pay close attention to the road and traffic conditions.

Fifth Practice Test 191

問43 正誤 □□	右の標識のあるところでは，横断する歩行者や自転車が明らかにいなければそのまま通行することができる。
問44 正誤 □□	歩行者用道路を通行するときは，歩行者が通行しているときでも，特に徐行しなくてもよい。
問45 正誤 □□	原動機付自転車を運転するときは，決められた速度の範囲内で，道路や交通の状況，天候や視界などに応じ，安全な速度を選ぶべきである。
問46 正誤 □□	疲れ，心配事，病気などのときは，判断力が衰えたりするので，運転を控える。

問47 交差点で右折待ちのため止まっていたら，対向車がライトを点滅させました。どのようなことに注意して運転しますか？

(1) 正誤 □□ 右折方向の横断歩道がよく見えないので，交差点の中央付近まで進み，横断歩道全体の様子も確認して右折する。

(2) 正誤 □□ トラックは前方が渋滞しているため，進路を譲ってくれたので，待たせないようにすばやく右折する。

(3) 正誤 □□ トラックのかげから二輪車が直進してくるかもしれないので，その様子を見ながら徐行する。

問48 前方が渋滞しています。この場合，どのようなことに注意して運転しますか？

(1) 正誤 □□ 後続車があるので，そのまま交差点に入って停止する。

(2) 正誤 □□ 左側の車の進路の妨げにならないように，交差点の手前で停止する。

(3) 正誤 □□ 自車のほうが優先道路で，左側の車は一時停止すると思われるので，交差点の中で停止する。

Q43. T F ☐☐	At locations with the sign on the right, you may proceed without stopping if there are clearly no pedestrians or bicycles crossing.	
Q44. T F ☐☐	When traveling on pedestrian pathways, there's no specific requirement to reduce speed, even when pedestrians are passing.	
Q45. T F ☐☐	When operating a motorized bicycle, it's essential to choose a safe speed within the specified speed range, considering factors such as road conditions, traffic, weather, and visibility.	
Q46. T F ☐☐	When feeling tired, worried, or unwell, judgment may be impaired, so it's advisable to refrain from driving.	

Q47. When stopped at an intersection waiting to turn right and the oncoming vehicle flashes its lights, what should you pay attention to while driving?

(1) T F ☐☐ Since the pedestrian crossing in the right-turn direction is not clearly visible, proceed to the center of the intersection to get a better view of the entire pedestrian crossing before making the right turn.

(2) T F ☐☐ A truck has graciously yielded the right of way due to the congestion ahead, so make the right turn quickly to avoid delaying them.

(3) T F ☐☐ Be cautious and drive at reduced speed while observing the possibility of a motorcycle behind the truck coming straight.

Q48. When there is congestion ahead, what should you be mindful of while driving?

(1) T F ☐☐ Since there are vehicles behind, proceed into the intersection and come to a stop.

(2) T F ☐☐ Stop before the intersection to avoid obstructing the path of the vehicle on the left.

(3) T F ☐☐ As my vehicle has the right of way and the vehicle on the left is likely to stop, I'll stop in the intersection.

Fifth Practice Test

| 第5回 解答と解説 | 間違えたら赤シートを当てて，覚えておきたいポイントを再チェックしよう！ |

◆・・・ひっかけ問題　　★・・・重要な問題

問1 ★正	片側2車線の道路の交差点であれば，標識の指定が無い場合，**小回りの右折方法**をとる。
問2 誤	二輪車のエンジンを切って押している場合は，**歩行者**と扱われる。なので，歩行者用信号に従う。
問3 ★誤	路面が乾燥しているときは前輪ブレーキを，路面が滑りやすいときは後輪ブレーキを少し強めにかける。
問4 ◆誤	後続車がいなくても，**合図**はする。
問5 誤	**高速道路**は原動機付自転車は通行することができない。
問6 ◆誤	乗降のため停車している通学通園バスのそばを通るときは，**徐行して安全確認**しなければならない。
問7 正	青色の左折の矢印信号がある交差点では，**左折**できる。
問8 正	原動機付自転車は**普通車**の仮免許では運転できない。
問9 誤	この標識は「安全地帯」を表している。
問10 誤	原付免許を取得して1年間は初心運転者期間といい，この間に違反をして一定の基準に達した人は初心運転者講習を受ける。
問11 誤	前方も注意するが，**バックミラー**などでも**周囲の状況**に目をくばる。
問12 誤	この標識は「**立ち入り禁止部分**」なので入れない。
問13 ◆誤	進路変更をしない場合は，合図をしてはいけない。

The Fifth Practice Test Answers and Explanations	If you make a mistake, use a red sheet to cover it and re-check the points you want to remember!

◆ · · · Tricky question　　★ · · · Important question

Q1. ★ T	If there are no specified instructions at intersections on two-lane roads, make **a tight right turn**.
Q2. F	When walking with a motorcycle with the engine off, you're considered a pedestrian, so follow traffic signals for **pedestrians**.
Q3. ★ F	Apply front wheel brakes when the road is dry; apply rear wheel brakes slightly stronger when the road is slippery.
Q4. ◆ F	Even if there are no following vehicles, **signaling** is necessary.
Q5. F	Motorcycles are not allowed on **highways**.
Q6. ◆ F	When passing by a school bus stopped for boarding and alighting, **reduce speed** and **perform safety checks**.
Q7. T	At intersections with a green left turn arrow signal, you can **turn left**.
Q8. T	You cannot operate a motorcycle with **a standard motor vehicle** provisional license.
Q9. F	This sign indicates a "safety zone."
Q10. F	During the first year after obtaining a motorized bicycle license, it's considered a novice driving period. Those who commit violations during this time must undergo novice driver training.
Q11. F	While paying attention to the front, also monitor **the surrounding conditions** using **rearview mirrors**.
Q12. F	This sign indicates a "**No Entry**" zone, so do not enter if you do not intend to change lanes.
Q13. ◆ F	If you are not changing lanes, you should not signal.

The Fifth Practice Test Answers and Explanations　195

問14 ★正	目がくらまないように，まぶしいときはやや左前方に視線を移すとよい。
問15 ★正	追越しに十分な余地がないときは，できる限り**左に寄って**進路を譲る。
問16 ◆誤	雨にぬれた道路は，制動距離が長くなるとともにスリップにも注意しなければならないので，急ブレーキは危険である。
問17 誤	幅の広い道路で右折するときは，**一気に移動せずに**徐々に中央寄りの車線に移っていく。
問18 正	右左折時は**徐行**しなければならない。
問19 誤	マフラーの改造の有無にかかわらず，他人の迷惑になるような騒音を出してはならない。
問20 ★誤	警察官が頭上に灯火を上げているとき，その警察官の身体の正面に平行する交通については，**黄色の信号**と同じ意味である。
問21 正	車の出入口から**3メートル**以内は駐車禁止である。
問22 正	「**転回禁止**」の標示である。
問23 誤	路側帯は**歩行者用**の通路なので，原動機付自転車は通行することができない。
問24 ◆誤	運行時間中に限り，バスの停留所の標示板（柱）から**10メートル**以内の場所では，停車も駐車も禁止。
問25 正	原動機付自転車の法定最高速度は時速**30キロメートル**毎時である。
問26 正	雨の降り始めの舗装道路や工事現場の鉄板やマンホールのふたなど，滑りやすいので注意して運転する。
問27 誤	この標識は「**危険物積載車両通行止め**」を表しているので，火薬類，爆発物，毒物，劇物などの危険物を積載する車は通行できない。

Q14. ★ T	To avoid being dazzled, it's advisable to shift your gaze slightly to the left when it's too bright.
Q15. ★ T	When there isn't enough space for overtaking, it's important to yield by moving as **far left** as possible.
Q16. ◆ F	On rainy roads, where braking distances are longer and there's a risk of slipping, sudden braking is dangerous and should be avoided.
Q17. F	When making a righet turn on a wide road, but gradually move into the center lane, **do not move allat once**.
Q18. T	It's necessary to **reduce speed** when making turns.
Q19. F	Regardless of whether the muffler has been modified, making excessive noise that bothers others is not permitted.
Q20. ★ F	When a police officer raises a light overhead, it signifies the same as **a yellow traffic signal** for traffic parallel to the officer's body.
Q21. T	Parking is prohibited within **3meters** of vehicle entrances and exits.
Q22. T	This sign indicates "**No U-turn**."
Q23. F	Since the side strip is designated for **pedestrian** use, motorized bicycles are not allowed to pass through.
Q24. ◆ F	During operating hours, parking and stopping are prohibited within **10 meters** of a bus stop sign (pole).
Q25. T	The legal maximum speed limit for motorized bicycles is **30 kilometers** per hour.
Q26. T	Be cautious when traversing slippery surfaces like newly wet roads or construction site iron plates and manhole covers.
Q27. F	This sign indicates "**No entry for vehicles carrying dangerous goods**," meaning vehicles transporting explosives, hazardous chemicals, toxins, or other dangerous materials are not allowed to pass.

問28 ◆誤	徐行する必要はないが，カーブの手前の直線部分であらかじめ**速度を落として**，安全な速度で通行する。
問29 ★誤	信号機のない踏切では，必ず**一時停止**する。
問30 誤	夜間は，できる限り視線を先のほうへ向け，いち早く前方の障害物を見つけられるようにする。
問31 誤	車の間をぬって走ったり，ジグザグ運転は危険なのでしてはいけない。
問32 ★誤	大地震で避難するときは，誰でも移動できるようにキーはつけっぱなしでよい。
問33 正	この標示は「**普通自転車の交差点進入禁止**」を表している。
問34 ★正	安全に停止できない場合は，そのまま進むことができる。
問35 正	水がたまっているところで，歩行者などのそばを通るときは，泥水がかかるおそれがあるので，速度を落として通行する。
問36 正	坂の頂上付近は**駐停車禁**止である。
問37 ★正	原動機付自転車の積載の重量限度は，**30キログラム**である。
問38 ★誤	車を離れるときは，短時間でもエンジンを止めなければならない。
問39 正	進路を変更するときは，周りの状況をしっかり把握する。
問40 正	この標識は「**前方優先道路**」を表している。
問41 ◆誤	前の車が交差点や踏切の手前で停止や徐行しているときは，その前に割り込んだり，横切らない。

Q28. ◆ F	There is no need to reduce speed, but it's advisable **to reduce speed** before reaching the straight section before a curve and proceed at a safe speed.
Q29. ★ F	At railway crossings without traffic signals, always **come to a stop**.
Q30. F	During nighttime, it's essential to focus your gaze as far ahead as possible to detect obstacles early.
Q31. F	We should avoid driving between cars or zigzag driving as it is dangerous.
Q32. ★ F	During evacuation in a major earthquake, it's permissible to leave the key in the ignition to allow anyone to move the vehicle.
Q33. T	This sign indicates "**No entry for standard bicycles at intersections**. "
Q34. ★ T	If unable to stop safely, you may continue straight ahead.
Q35. T	When passing through areas with pooled water near pedestrians, reduce speed to avoid splashing them with mud.
Q36. T	**Parking is prohibited** near the top of a hill.
Q37. ★ T	The weight limit for cargo on a motorized bicycle is **30 kilograms**.
Q38. ★ F	Even for a short duration, the engine must be turned off when leaving the vehicle.
Q39. T	When changing lanes, it's crucial to thoroughly assess the surrounding conditions.
Q40. T	This sign indicates "**Priority Road Ahead**. "
Q41. ◆ F	When the preceding vehicle is reducing speed or stopping before an intersection or railway crossing, avoid cutting in front of it or crossing in front.

The Fifth Practice Test Answers and Explanations   199

問42 正	交通規則を守って，そのときの状況に応じられるように注意する。
問43 ★正	この標識は「横断歩道・自転車横断帯」を表していて，横断する歩行者，自転車がいるときは，一時停止して，あきらかにいない場合はそのまま通行できる。
問44 誤	歩行者用道路を通行するときは，歩行者に十分注意して徐行する。
問45 正	道路や交通の状況，天候など，その場の状況を把握して安全な速度で運転する。
問46 正	身体の調子が悪いときは，運転は控える。

問47　歩行者とトラックのかげに要注意！！
自車に進路を譲るサインで，ライトの点滅があります。この場合でも，トラックのかげに注意して安全確認してから右折しよう。

(1)　正　歩行者の横断に，注意する。
(2)　誤　二輪車がトラックのかげから，出てくるおそれがある。
(3)　正　二輪車が出てくるかもしれないので注意し，徐行して右折する。

問48　左側の車と前方の渋滞に要注意！！
渋滞していると，交差点内で停止してしまうおそれがあるので，他車の交通の妨げにならないように交差点内を避けよう。

(1)　誤　後続車があっても，交差点内での停止は禁止。
(2)　正　交差点の手前で停止すれば，左側の車の進行を妨げないのでよい。
(3)　誤　交差点内で停止すると，左側の車の妨げになるので，交差点の手前で停止する。

Q42. T	Observe traffic regulations and remain attentive to adapt to the current situation.
Q43. ★ T	This sign indicates "**Pedestrian Crossing/ Bicycle Crossing**," and when pedestrians or bicycles are crossing, it's necessary to come to **a temporary stop**; if there are none visibly present, **you may proceed**.
Q44. F	When traveling on **pedestrian roads**, proceed **with caution** and pay close attention to pedestrians.
Q45. T	Drive at a safe speed by assessing the current situation, including road and traffic conditions, as well as the weather.
Q46. T	If feeling unwell physically, refrain from driving.

Q47.　Watch out for **pedestrians** and the **blind** spots of trucks!
Even if there's **a sign indicating** that the car is yielding the right of way to you with flashing lights, be sure **to check** for trucks in your blind spots and confirm safety before making a right turn.

（1）　T　Pay attention to pedestrian crossings.
（2）　F　There is a risk of motorcycles emerging from **the back** of trucks.
（3）　T　Be cautious as motorcycles may emerge, so **reduce speed** and make a right turn.

Q48.　Pay attention to the cars on **the left** and the **traffic jam** ahead.
There is a risk of **stopping** in the intersection when there is traffic congestion, so try to avoid being an obstruction to other traffic by staying out of the intersection.

（1）　F　It is prohibited **to stop** in the intersection, even if there are trailing vehicles.
（2）　T　It is acceptable to stop before the intersection, as it does not obstruct the progress of the vehicles on **the left**.
（3）　F　Stopping in the intersection **obstructs** the progress of vehicles on the left, so it is preferable to stop before entering the intersection.

The Fifth Practice Test Answers and Explanations　201

第6回 模擬試験	問1～問46までは各1点，問47，48は全て正解して各2点。制限時間30分，50点中45点以上で合格

●次の問題で正しいものは「正」，誤りのものは「誤」の枠をぬりつぶして答えなさい。

問1 正誤 □□	交通量が少ないときは，他の歩行者や車に迷惑をかけることはないので，自分の都合だけを考えて運転してもよい。
問2 正誤 □□	車の内輪差は曲がるときに徐行をすれば生じない。
問3 正誤 □□	信号待ちで原動機付自転車が停止している状態でも，厳密には運転中に当たるので，携帯電話は使用しない。
問4 正誤 □□	長い下り坂では，ガソリンを節約するため，エンジンを止め，ギアをニュートラルにして，ブレーキを使用した方がよい。
問5 正誤 □□	原付免許で運転できる車は，原動機付自転車だけである。
問6 正誤 □□	交通整理の行われていない，道幅が同じような交差点にさしかかった場合，車は路面電車の通行を妨げてはならない。
問7 正誤 □□	消火栓，消防水利の標識がある場所や，消防用防火水槽の取入口から5メートル以内の場所では，駐車も停車もしてはならない。
問8 正誤 □□	原動機付自転車の乗車定員は2人である。
問9 正誤 □□	右の標識のある交差点では，直進してその交差点を通過してはならない。
問10 正誤 □□	警察官の手信号で，両腕を水平にあげた状態に対面した車は，停止位置を越えて進行することはできない。
問11 正誤 □□	トンネルの中では，対向車に注意を与えるため，右側の方向指示器を作動させたまま走行したほうがよい。
問12 正誤 □□	右の標識のある交通整理が行われている交差点を原動機付自転車で右折しようとするときは，十分手前から徐々に中央寄りの車線に移るようにする。
問13 正誤 □□	黄色の線で区画されている車両通行帯でも，後続車がない場合は，その線を越えて進路変更してもよい。

Sixth Practice Test	Questions 1 to 46 are worth 1 point each, while questions 47 and 48 are worth 2 points each if all answers are correct. The time limit is 30 minutes. To pass, you need 45 points out of 50.

● Fill in the box marked 'T' for the correct answers and 'F' for the incorrect ones in the following questions.

Q1. T F ☐☐	When traffic is light, it is acceptable to drive only considering one's own convenience because you don't bother other pedestrians or vehicles.
Q2. T F ☐☐	The inner wheel difference in a car does not occur if you reduce speed when turning.
Q3. T F ☐☐	Even when a motorized bicycle is stopped at a signal, it is technically considered to be driving, so the use of a mobile phone is prohibited.
Q4. T F ☐☐	To save gasoline on long downhill slopes, it is advisable to turn off the engine, shift to neutral, and use the brakes.
Q5. T F ☐☐	Vehicles that can be driven with a motorized bicycle license are only motorized bicycle.
Q6. T F ☐☐	When approaching an intersection where traffic control is not conducted and the road width is similar, cars must not obstruct the passage of streetcars.
Q7. T F ☐☐	Parking or stopping is prohibited within 5 meters of fire hydrants, fire water signs, or fire-fighting water tank access points.
Q8. T F ☐☐	The maximum passenger capacity of a motorized bicycle is two people.
Q9. T F ☐☐	At intersections with the right sign, vehicles must not proceed straight through the intersection.
Q10. T F ☐☐	When a police officer signals with both arms raised horizontally, vehicles facing them must not proceed beyond the stopping position.
Q11. T F ☐☐	In tunnels, it is advisable to keep the right turn signal on to alert oncoming vehicles.
Q12. T F ☐☐	When attempting to turn right at an intersection where traffic control is being conducted with a motorized bicycle, gradually move towards the center lane far enough infront before the turn.
Q13. T F ☐☐	Even in vehicle passing lanes delineated by yellow lines, if there are no following vehicles, it is permitted to change lanes beyond those lines.

Sixth Practice Test 203

問14 正誤 □□	トンネルの中や霧などで視界が悪いときに，右側の方向指示器を出して走行すると，後続車の判断を誤らせ，迷惑になるのでしてはならない。
問15 正誤 □□	二輪車を運転するときは，工事用安全帽をかぶれば，乗車用ヘルメットの代わりにすることができる。
問16 正誤 □□	行き違いができないような狭い坂道では，原則として下りの車が上りの車に道を譲る。
問17 正誤 □□	路線バス等優先通行帯は，路線バスのほか軽車両だけが通行できる。
問18 正誤 □□	一方通行の道路では，道路の中央から右側部分にはみ出して通行することができない。
問19 正誤 □□	放置車両確認標章を取り付けられた車の使用者は，放置違反金の納付を命ぜられることがある。
問20 正誤 □□	夜間は原動機付自転車はほかの運転者から見えにくいので，なるべく目につきやすい服装にするとよい。
問21 正誤 □□	右の標示板がある場合は，信号機の信号に関係なく左折できる。  ←
問22 正誤 □□	前の車が交差点や踏切の手前で徐行しているときは，その前を横切ってはならないが，停止しているときは，その前を横切ってもよい。
問23 正誤 □□	夜間走行中の前照灯は，下向きに切り替えると前方の視界が悪くなって危険なので，常に上向きにしておくべきである。
問24 正誤 □□	人の健康や生活環境に害を与える自動車の排気ガスは，速度や積載の超過とは関係がない。
問25 正誤 □□	車両通行帯のない道路では，速度の速い車は，原則として道路の中央寄りの部分を通行しなければならない。
問26 正誤 □□	下り坂のカーブに，右の図の標示があるときは，対向車に注意しながら道路の右側部分にはみ出すことができる。 
問27 正誤 □□	ブレーキレバーやブレーキペダルのあそびが調整されていない車は，速度を落として運転するとよい。
問28 正誤 □□	同一方向に進行しながら進路を右に変える場合，後続車がいなければ合図をする必要はない。

Q14. T F ☐☐	In tunnels or in foggy conditions where visibility is poor, it is prohibited to use the right turn signal while driving straight ahead, as it may mislead following vehicles and cause inconvenience.
Q15. T F ☐☐	When driving a motorcycle, wearing a construction safety helmet can serve as a substitute for a motorcycle helmet.
Q16. T F ☐☐	On narrow slopes where passing is impossible, the general rule is for downhill vehicles to yield to uphill vehicles.
Q17. T F ☐☐	Priority lanes for buses only allow passage for buses and light road vehicles.
Q18. T F ☐☐	On one-way roads, it is not permitted to drive protruding into the right side from the center of the road.
Q19. T F ☐☐	Users of vehicles fitted with an abandoned vehicle confirmation sign may be ordered to pay a fine for abandonment violations.
Q20. T F ☐☐	Since motorized bicycles are less visible to other drivers at night, it is advisable to wear clothing that is easily visible.
Q21. T F ☐☐	When there is a this sign, left turns are permitted regardless of traffic signals. 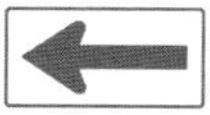
Q22. T F ☐☐	When the vehicle in front is reducing speed before an intersection or railroad crossing, it is prohibited to cross in front of it, but it is permitted to do so when it is stopped.
Q23. T F ☐☐	Adjusting the headlights downward during nighttime driving is dangerous as it impairs visibility, so they should always be kept in the upward position.
Q24. T F ☐☐	Vehicle exhaust gases that harm human health or the environment are not related to speeding or overloading.
Q25. T F ☐☐	On roads without vehicle passing lanes, faster vehicles should generally travel towards the center of the road.
Q26. T F ☐☐	When there is a downhill slope with the sign shown on the right, it is permitted to drive protruding into the right side of the road while being cautious of oncoming traffic. 
Q27. T F ☐☐	It is advisable to reduce speed when braking levers or pedals have excessive play.
Q28. T F ☐☐	When changing lanes to the right while traveling in the same direction, signaling is not necessary if there are no following vehicles.

問29 正誤 ☐☐	信号機のあるところでは，前方の信号に従うべきであって，横の信号が赤になったからといって発進してはならない。
問30 正誤 ☐☐	二輪車でカーブを曲がるときは，ハンドルを切るのではなく，車体を傾けることによって自然に曲がるような要領で行うのがよい。
問31 正誤 ☐☐	二輪車のマフラーは，取り外しても事故の原因にはならないので，取り外して運転してもかまわない。
問32 正誤 ☐☐	右の図の標識は，この先が行き止まりであることを表している。 
問33 正誤 ☐☐	二輪車のブレーキのかけ方には，ブレーキレバーを使う場合，ブレーキペダルを使う場合，エンジンブレーキを使う場合の3種類がある。
問34 正誤 ☐☐	標識には本標識と補助標識があり，本標識は規制標識，指示標識，警戒標識の3種類だけである。
問35 正誤 ☐☐	原動機付自転車は高速自動車国道は走れないが，自動車専用道路は通行できる。
問36 正誤 ☐☐	不必要な急発進や急ブレーキ，空ぶかしは危険ばかりでなく，交通公害のもととなる。
問37 正誤 ☐☐	交通事故を起こしたときは，負傷者の救護より先に警察や家族に電話で報告しなければならない。
問38 正誤 ☐☐	道路に面した場所に出入りするために歩道を横切る場合は，歩行者がいなければ徐行して通行することができる。
問39 正誤 ☐☐	右の標識がある場所でも，警察官の手信号に従うときは，一時停止しなくてもよい。 
問40 正誤 ☐☐	横断歩道を歩行者が横断していたが，車を見て立ち止まったので，そのまま通過した。
問41 正誤 ☐☐	補助標識は本標識の意味を補足するもので，すべて本標識の下に取り付けられる。
問42 正誤 ☐☐	右の標識のある場所を通る車は，必ず警音器を鳴らさないといけない。 
問43 正誤 ☐☐	前の車に続いて踏切を通過するときは，一時停止しなくてよい。

Q29. T F	At signalized intersections, it is necessary to obey the signal ahead, and one should not start moving just because the side signal has turned red.
Q30. T F	When turning on a motorcycle, it is advisable to lean the body naturally into the curve rather than just turning the handlebars.
Q31. T F	Removing the muffler from a motorcycle does not cause an accident, so it is permitted to drive without it.
Q32. T F	The sign depicted on the right indicates that the road ahead is a dead end. 
Q33. T F	There are three ways to apply the brakes on a motorcycle: using the brake lever, using the brake pedal, and using the engine brake.
Q34. T F	Signs can be divided into main signs and supplementary signs, with main signs being regulatory signs, directional signs, and warning signs.
Q35. T F	While motorized bicycles cannot travel on expressways, they are permitted on roads designated for automobiles only.
Q36. T F	Unnecessary rapid acceleration, sudden braking, and excessive engine revving are not only dangerous but also contribute to traffic pollution.
Q37. T F	In the event of a traffic accident, reporting to the police or family members by phone takes precedence over providing aid to the injured.
Q38. T F	When crossing a sidewalk to enter or exit a side strip, one may drive at a reduced speed if there are no pedestrians.
Q39. T F	Even at locations with the sign depicted on the right, it is not necessary to come to a stop when following a police officer's hand signals. 
Q40. T F	If pedestrians are crossing at a pedestrian crossing and a vehicle stops to let them pass, other vehicles may proceed without stopping.
Q41. T F	Supplementary signs provide additional information to the main sign and are always installed below the main sign.
Q42. T F	Vehicles passing through a location with the sign shown on the right must always sound their horn. 
Q43. T F	When following a preceding vehicle through a railway crossing, it is not necessary to come to a stop.

Sixth Practice Test　207

問44 正誤 □□	二輪車に乗るときは，たとえ暑い季節でも，身体の露出が少なくなるような服装をしたほうがよい。
問45 正誤 □□	転回や右折をするときは，それらの行為をしようとする約3秒前に，合図をしなければならない。
問46 正誤 □□	夜間走行中，対向車のライトがまぶしい場合は，ライトを直視し，目を光に慣れさせることが大切である。

問47　時速30キロメートルで進行しています。どのようなことに注意して運転しますか？

(1) 正誤 □□　対向車が通過するまで，駐車車両の後方で一時停止して道を譲る。

(2) 正誤 □□　駐車車両は，急に発進するかもしれないので，速度を落として車の様子を見る。

(3) 正誤 □□　先に道路の右側部分にはみ出せば，対向車は道を譲ってくれると思うので，加速して駐車車両の側方を通過する。

問48　時速10キロメートルで進行しています。交差点を右折するときは，どのようなことに注意して運転しますか？

(1) 正誤 □□　トラックのかげから二輪車が直進してくるかもしれないので，トラックが直進するのを待ち，前方を確認してから右折する。

(2) 正誤 □□　交差点内の車が右折したら，すばやく右折する。

(3) 正誤 □□　右折方向の横断歩道上には歩行者が通行しているので，これを妨げないようにして右折する。

Q44. T F ☐☐	When riding a motorcycle, it is advisable to wear clothing that minimizes exposure to the body, even in hot weather.
Q45. T F ☐☐	When preparing to make a U-turn or right turn, it is necessary to signal approximately 3 seconds before initiating these actions.
Q46. T F ☐☐	During nighttime riding, if the headlights of oncoming vehicles are bright, it is important to stare directly at the lights and allow your eyes to adjust to the light.

Q47. When driving at a speed of 30 kilometers per hour, what should you pay attention to?

(1) T F ☐☐ You will stop temporarily behind parked vehicles until oncoming traffic passes.

(2) T F ☐☐ Parked vehicles may start suddenly, so you will reduce speed and observe their behavior.

(3) T F ☐☐ If you protrude into the right side of the road, oncoming vehicles are likely to yield, so you will accelerate and pass alongside the parked vehicles.

Q48. When traveling at a speed of 10 kilometers per hour, what should you be careful about when making a right turn at an intersection?

(1) T F ☐☐ Since a motorcycle might be coming straight from the shadow of the truck, you will wait for the truck to go straight, check ahead, and then make a right turn.

(2) T F ☐☐ When the cars inside the intersection make a right turn, you will quickly make my right turn.

(3) T F ☐☐ As pedestrians are crossing on the pedestrian crossing in the direction of the right turn, you will make the right turn without obstructing them.

| 第6回
解答と解説 | 間違えたら赤シートを当てて，覚えておきたいポイントを再チェックしよう！ |

◆・・・ひっかけ問題　　★・・・重要な問題

問1 ★誤	交通量が少なくても，自分本位の運転をしてはいけない。
問2 誤	カーブを曲がるときは，必ず内輪差が生じる。
問3 正	運転前に電源を切ったり，ドライブモードにしておく。
問4 誤	長い下り坂で，頻繁にブレーキを使用すると，急にブレーキがきかなくなることがある。
問5 正	原付免許では原動機付自転車のみ運転できる。
問6 ◆正	左右関係なく，交通整理の行われていない道幅が同じような交差点では，**路面電車**が優先する。
問7 誤	駐車は禁止されているが，**停車**は禁止されていない。
問8 誤	原動機付自転車の乗車定員は1人である。
問9 正	「**指定方向外進行禁止**」の標識なので，直進できない。
問10 正	警察官の手信号で，両腕を横に水平にあげた状態に対面した車は，**停止位置**を越えて進行することはできない。
問11 ◆誤	進路変更などしないのに合図してはならない。
問12 ★誤	車両通行帯が3車線の道路の交差点での右折は，原則として**二段階右折**する。
問13 ★誤	後続車がいなくても，黄色の線で区画された車両通行帯は進路変更禁止である。

210　第6回模擬試験　解答と解説

The Sixth Practice Test Answers and Explanations	If you make a mistake, use a red sheet to cover it and re-check the points you want to remember!

◆ · · · Tricky question   ★ · · · Important question

Q1. ★ F	Even if there is little traffic, one should not act selfishly.
Q2. F	When turning on a curve, a wheel difference always occurs.
Q3. T	Before driving, turn off the power or keep it in drive mode.
Q4. F	When using brakes frequently on long downhill slopes, there may be a sudden loss of braking power.
Q5. T	With a motorized bicycle license, only motorized bicycles can be driven.
Q6. ◆ T	Regardless of direction, at intersections where the width of the road is the same and traffic control is not conducted, **streetcars** have priority.
Q7. F	Parking is prohibited, but **stopping** is not.
Q8. F	The riding capacity of a motorized bicycle is one person.
Q9. T	It is a "**prohibited direction of travel**" sign, so you cannot go straight.
Q10. T	When a police officer signals with both arms raised horizontally facing each other, vehicles cannot proceed beyond **the stopping position**.
Q11. ◆ F	Signaling should not be done if there is no change in course.
Q12. ★ F	For three-lane roads with vehicle lanes, right turns at intersections are generally made **in two stages**.
Q13. ★ F	Even if there are no following vehicles, changing lanes is prohibited in lanes delineated by yellow lines.

The Sixth Practice Test Answers and Explanations   211

問14 正	右折しないのに，方向指示器を出してはいけない。
問15 誤	工事用安全帽は乗車用ヘルメットではないので，PS（c）マークかJISマークのついたヘルメットをかぶる。
問16 正	上り坂の方が発進が難しいので，原則として下りの車が上りの車に道を譲る。
問17 ★誤	路線バス等優先通行帯は，自動車や原動機付自転車も通行できる。
問18 誤	**一方通行**の道路では，中央より右側も通行できる。
問19 正	運転者が反則金の納付など行わなかった場合は，使用者に放置違反金を命ぜられる場合もある。
問20 正	夜間は暗くてほかの運転者から見えにくいので，目につきやすい服装がよい。
問21 正	信号機の信号に関係なく，左折できる。
問22 誤	前の車が交差点や踏切などで停止や徐行しているときは，割り込んだり横切ったりはしてはいけない。
問23 ◆誤	交通量の多い市街地や対向車があるときなどでは，前照灯を**下向き**に切り替えて運転する。
問24 誤	速度超過や過積載は交通公害の原因になる。
問25 誤	速度に関係なく，追越しなど以外，道路の**左**に寄り通行しなければならない。
問26 正	右側通行を示す標示で，対向車に注意しながら右側部分に**はみ出して**通行することができる。
問27 誤	ブレーキ装置のあそびが整備されてない車は**整備不良**となり，運転してはならない。

Q14. T	Do not activate the turn signal when not turning right.
Q15. F	Construction safety helmets are not the same as riding helmets, so wear a helmet with a PS (c) mark or JIS mark.
Q16. T	As uphill starts are more difficult, generally, downhill vehicles should yield to uphill ones.
Q17. ★ F	Priority lanes for route buses can also be used by cars and motorized bicycles.
Q18. F	On **one- way** roads, vehicles can also travel on the right side of the center.
Q19. T	If a driver does not pay a fine, the user may be required to pay a parking violation fine.
Q20. T	During the night, when it's dark and drivers are less visible to others, it's good to wear conspicuous clothing that is easily visible.
Q21. T	Left turns can be made regardless of traffic signals.
Q22. F	When the preceding vehicle is stopped or reducing speed at intersections or railroad crossings, one should not cut in or cross.
Q23. ◆ F	In busy urban areas or when there are oncoming vehicles, switch the headlights to low beam while driving.
Q24. F	Speeding and overloading contribute to traffic pollution.
Q25. F	Regardless of speed, except for passing, vehicles must travel close to the **left** side of the road.
Q26. T	With a sign indicating right- side passage, one can drive **protruding** to the right side while being cautious of oncoming traffic.
Q27. F	Vehicles with poorly adjusted brake play are considered **poorly maintained** and should not be driven.

The Sixth Practice Test Answers and Explanations 213

問28 ★誤	後続車がいる，いないにかかわらず合図しなければならない。
問29 正	前方の信号に従わなければならない。
問30 正	ハンドルだけで曲がろうとすると転倒するかもしれないので，車体を傾けて自然に曲がるようにする。
問31 誤	マフラーを取り外すと騒音が大きくなり，周囲に迷惑がかかる。
問32 誤	「その他の危険」の標識のため，行き止まりを意味する標識ではない。
問33 ★正	二輪車のブレーキのかけ方には，ブレーキレバー，ブレーキペダル，エンジンブレーキを使う場合の3種類がある。
問34 誤	本標識は，規制標識，指示標識，警戒標識のほかに案内標識の4種類がある。
問35 ★誤	原動機付自転車は高速自動車国道や自動車専用道路を走ることはできない。
問36 正	急発進や急ブレーキなど危険運転であり，交通公害にもなる。
問37 ★誤	交通事故を起こしたら，事故の続発も防ぐために負傷者の救護を行う。
問38 誤	歩行者の有無にかかわらず，歩道の手前で一時停止して安全確認をする。
問39 正	「一時停止」の標識だが，警察官の手信号に従う場合はそちらの方が優先する。
問40 誤	歩行者が横断歩道を横断しているときは，一時停止して，歩行者の横断を妨げない。
問41 ◆誤	「終わり」の標識のように，本標識の上にも取り付けられる場合もある。

Q28. ★ F	**Signal** must be given regardless of whether there are following vehicles or not.
Q29. T	One must follow the signal ahead.
Q30. T	Trying to turn with just the handlebars may cause **a fall**, so lean the body to **naturally** make the turn.
Q31. F	Removing the muffler increases noise and causes inconvenience to the surrounding area.
Q32. F	This sign indicates "**other hazards**" and does not mean a dead end.
Q33. ★ T	There are **three ways** to apply brakes on a motorcycle: using the brake lever, brake pedal, or engine brake.
Q34. F	Besides regulatory signs, instructional signs, and warning signs, there are also **guidance signs** among these four types of signs.
Q35. ★ F	Motorized bicycles cannot travel on expressways or express lanes for automobiles.
Q36. T	Sudden acceleration or braking constitutes dangerous driving and contributes to traffic pollution.
Q37. ★ F	In the event of a traffic accident, providing first aid to the injured is essential to prevent further accidents.
Q38. F	Regardless of pedestrians, **stop temporarily** before the sidewalk and perform safety checks.
Q39. T	Although it is a "stop" sign, when following a police officer's hand signal, it takes precedence.
Q40. F	When pedestrians are crossing the pedestrian crossing, **stop temporarily** and do not obstruct their crossing.
Q41. ◆ F	Similar to the "end" sign, additional signs can also be mounted above this sign.

The Sixth Practice Test Answers and Explanations 215

問42 正	この標識は「警笛鳴らせ」なので，この標識がある場所では，必ず警音器を鳴らして自車の接近を知らせる。
問43 ★誤	踏切を前の車に続いて通過する場合でも，**一時停止**して安全確認をする。
問44 正	転倒して怪我をしてしまうことを考えて，長袖や長ズボンなどの服装で，プロテクターを装着するとよい。
問45 誤	転回や右折の合図は，**30メートル**手前の地点に達したときに行う。
問46 ◆誤	ライトを直視しないで，視線をやや左前方に向けて目がくらまないようにする。

問47　駐車車両と対向車に要注意！！
駐車している車がいきなり動き出すかもしれないので注意して，対向車にも衝突しないように気をつけよう。

(1)　正　障害物などあるときは，一時停止などして**対向車**に道を譲る。
(2)　正　**駐車車両**がいきなり発進するかもしれないので，減速して様子をみる。
(3)　誤　右側部分にはみ出すと，**対向車**と衝突するおそれがある。

問48　トラックのかげと歩行者に要注意！！
トラックのかげから出てくるかもしれない車などに注意して，横断歩道を渡っている歩行者の通行を妨げないようにしよう。

(1)　正　トラックが直進するのを待ち，前方の**安全確認**をして右折する。
(2)　誤　トラックの**かげ**から他の車が直進して，衝突するおそれがある。
(3)　正　横断歩道を渡っている**歩行者の通行**を妨げないように右折する。

Q42. T	This sign means "**Sound Horn**", so at locations where this sign is present, always sound the horn to alert others of your approach.
Q43. ★ F	Even when following the preceding vehicle through a railroad crossing, **stop temporarily** and perform safety checks.
Q44. T	To prevent injuries from falling, it is advisable to wear protective gear such as long-sleeved shirts and pants.
Q45. F	Signals for turning or making a right turn should be given when reaching a point **30 meters** before the turn.
Q46. ◆ F	Avoid looking directly at the headlights and instead, shift your gaze slightly to the left front to prevent dazzling.

Q47. Be cautious of **parked vehicles** and **oncoming cars!!**
Be vigilant because parked cars may suddenly start moving, and be careful not to collide with oncoming vehicles.

(1) T When there are obstacles, etc., stop temporarily and give way to **oncoming vehicles**.

(2) T Since **parked vehicles** may suddenly start moving, reduce speed and observe the situation.

(3) F There is a risk of collision with **oncoming vehicles** if you protrude into the right side.

Q48. Be cautious of **behind** the trucks and **the pedestrians!**
Be mindful of vehicles that may emerge from the blind spots of trucks and ensure that you do not obstruct pedestrians crossing the crosswalk.

(1) T Wait for the truck to proceed straight and then perform **a safety check** ahead before making a right turn.

(2) F There is a risk of collision if another vehicle proceeds straight from the **blind spot** of the truck.

(3) T Make a right turn without **obstructing pedestrians** crossing the crosswalk.

第7回 模擬試験	問1〜問46までは各1点，問47，48は全て正解して各2点。制限時間30分，50点中45点以上で合格

●次の問題で正しいものは「正」，誤りのものは「誤」の枠をぬりつぶして答えなさい。

問1 正誤 □□	原動機付自転車は，車両通行帯のない道路では，道路の中央寄りを通行しなければならない。
問2 正誤 □□	チェーンの中央部分を指で押したところ，20ミリメートルぐらいのゆるみがあったので適当と判断し，そのまま運転した。
問3 正誤 □□	左側部分の道幅が6メートル未満の道路で，中央に黄色の線が引かれているところでも，右側部分にはみ出さなければ追い越ししてもよい。
問4 正誤 □□	原動機付自転車の法定最高速度は，20キロメートル毎時である。
問5 正誤 □□	運転免許試験に合格すれば，免許証を交付される前に原動機付自転車を運転しても無免許運転ではない。
問6 正誤 □□	車輪のガタは，後輪よりも前輪のほうが運転に大きな影響を与える。
問7 正誤 □□	車を発進させるときは，バックミラーだけで後方を確認し，急発進させて車の流れの中に入ったほうがよい。
問8 正誤 □□	原動機付自転車のマフラーの破損は，運転に直接影響はないので，そのままにしておいてもよい。
問9 正誤 □□	右の標識のある道路で「原付を除く」の補助標識があれば，原動機付自転車はその道路を通行することができる。
問10 正誤 □□	原動機付自転車で前方の信号が青のときは，直進，左折，右折することができる。（二段階右折の場合を除く）
問11 正誤 □□	原動機付自転車を押して歩く場合は，すべて歩行者とみなされる。
問12 正誤 □□	右の標識のある通行帯は自動二輪車は通行できるが，原動機付自転車は通行できない。
問13 正誤 □□	環状交差点に進入するときは，必ず左折の合図を行わなければならない。

218 第7回模擬試験 問題

	Questions 1 to 46 are worth 1point each, while questions 47 and 48 are worth
Seventh Practice Test	2 points each if all answers are correct. The time limit is 30 minutes. To pass, you need 45 points out of 50.

● Fill in the box marked 'T' for the correct answers and 'F' for the incorrect ones in the following questions.

Q1. T F ☐☐	Motorized bicycles must travel closer to the center of the road on roads without vehicle lanes.
Q2. T F ☐☐	I judged the chain to be adequate as there was about 20 millimeters of slack when pressed in the middle with a finger, and continued driving.
Q3. T F ☐☐	On roads where the width of the left side is less than 6 meters and there is a yellow line in the center, it is permitted to overtake without protruding into the right side.
Q4. T F ☐☐	The statutory maximum speed limit of a motorized bicycle is 20 kilometers per hour.
Q5. T F ☐☐	Driving a motorized bicycle before being issued a license does not constitute driving without a license if you pass the driving test.
Q6. T F ☐☐	Wheel play has a greater impact on driving for the front wheels than the rear wheels.
Q7. T F ☐☐	When starting the car, it is advisable to only check the rearview mirror and accelerate quickly to merge into traffic.
Q8. T F ☐☐	Damage to the muffler of a motorized bicycle does not directly affect driving, so it can be left as is.
Q9. T F ☐☐	If there is an auxiliary sign indicating "原付を除く (Except Motorized Bicycles)" on a road with a sign shown on the right, motorized bicycles are allowed to travel on that road.
Q10. T F ☐☐	When the front signal is green on a motorized bicycle, you can go straight, turn left, or turn right (except for two-stage right turns).
Q11. T F ☐☐	When walking with a motorized bicycle, it is considered as pedestrian.
Q12. T F ☐☐	Motorcycles can travel in lanes with signs shown on the right, but motorized bicycles cannot.
Q13. T F ☐☐	When entering a roundabout, you must always signal left.

Seventh Practice Test **219**

問14 正誤 □□	みだりに車両通行帯を変えながら通行することは，後続車の迷惑となったり事故の原因にもなる。
問15 正誤 □□	身体の不自由な人を乗せた車いすを，健康な人が押して通行している場合は，一時停止や徐行をする必要はない。
問16 正誤 □□	原動機付自転車はいつでも自動車と同じ方法で右折することができる。
問17 正誤 □□	原動機付自転車ならば，一方通行となっている道路を逆方向へ進行することができる。
問18 正誤 □□	2本の白線で区画されている路側帯は，その幅が広いときに限り，中に入って駐停車することができる。
問19 正誤 □□	追越しが終わったら，すぐに追い越した車の前に入るのがよい。
問20 正誤 □□	転回の合図は右折の合図と同じである。（環状交差点での転回を除く）
問21 正誤 □□	右の標識のある場所ではハンドルをしっかりと握り注意して運転する。 
問22 正誤 □□	こう配の急な登り坂であっても，5分以内の荷物の積み下ろしならば，停車することができる。
問23 正誤 □□	路線バス等優先通行帯を走行中，バスが近づいてきたら原動機付自転車はそこから出なければならない。
問24 正誤 □□	道路の左寄り部分が工事中のときは，いつでも道路の中央から右側にはみ出して走行してもよい。
問25 正誤 □□	踏切の向こう側が混雑しているため，そのまま進むと踏切内で動きがとれなくなるおそれがあるときは，踏切に入ってはならない。
問26 正誤 □□	右の図の標示のある道路では，原動機付自転車は左側の通行帯を通行する。 
問27 正誤 □□	進路変更が終わった約3秒後に合図をやめた。

220　第7回模擬試験　問題

Q14. T F ☐☐	Changing lanes indiscriminately while driving can cause inconvenience to following vehicles and may also lead to accidents.
Q15. T F ☐☐	When a healthy person is pushing a wheelchair carrying a person with physical disabilities, there is no need to stop or reduce speed.
Q16. T F ☐☐	Motorized bicycles can make right turns anytime using the same method as cars.
Q17. T F ☐☐	Motorized bicycles are allowed to travel in the opposite direction on one-way streets.
Q18. T F ☐☐	Side strips divided by two white lines can be used for parking or stopping only when it is large in width.
Q19. T F ☐☐	After completing an overtaking maneuver, it is advisable to merge in front of the overtaken vehicle.
Q20. T F ☐☐	The signal for turning is the same as for making a right turn (except at roundabouts).
Q21. T F ☐☐	When there is a sign shown on the right, it is important to grip the handle firmly and drive attentively. 
Q22. T F ☐☐	Even on steep uphill slopes, stopping is permitted for unloading or loading cargo within 5minutes.
Q23. T F ☐☐	When traveling in a lane designated for priority use by buses and similar vehicles, motorized bicycles must get out of the lane when buses approaching.
Q24. T F ☐☐	When the left side of the road is under construction, it is permissible to drive protruding from the center to the right side of the road at any time.
Q25. T F ☐☐	If it is crowded beyond the railroad crossing, entering the crossing may result being unable to move in it, so do not enter.
Q26. T F ☐☐	On roads with the sign depicted on the right, motorized bicycles should travel in the left lane. 
Q27. T F ☐☐	The signal was discontinued approximately 3 seconds after completing the lane change.

問28 正誤 □□	道路の曲がり角付近では追越しが禁止されている。
問29 正誤 □□	原動機付自転車が，リヤカーでけん引するときの法定最高速度は，20キロメートル毎時である。
問30 正誤 □□	横断歩道と自転車横断帯は，横断するのが歩行者と自転車の違いだけで，原動機付自転車が通行する方法は変わらない。
問31 正誤 □□	ぬかるみのある場所では，低速ギアなどを使い速度を落として通行する。
問32 正誤 □□	右の標識が示されていたので，そのスピードで原動機付自転車を運転した。
問33 正誤 □□	交差点の中まで車両通行帯の線が引かれていても，優先道路の標識がなければ，優先道路ではない。
問34 正誤 □□	原動機付自転車が，上り坂の頂上付近で，徐行している原動機付自転車を追い越した。
問35 正誤 □□	踏切とその端から前後10メートル以内の場所は短時間であっても，停車することはできない。
問36 正誤 □□	大地震が発生したときは，機動力のある原動機付自転車に乗って避難する。
問37 正誤 □□	カーブを走行中にハンドルを右に切ると，バイクは左に倒れようとする。
問38 正誤 □□	原動機付自転車は路面電車が通行していないときなら，いつでも軌道敷地内を通行することができる。
問39 正誤 □□	右のイラストのように，警察官が手信号による交通整備を行っている場合，AとBは同じ意味である。
問40 正誤 □□	子どもが道路上で遊んでいたので，警音器を鳴らして注意させ，その横を通過した。
問41 正誤 □□	片側2車線の道路の交差点で信号機が青を標示しているときには，原動機付自転車は，左折や小回り右折をすることができる。
問42 正誤 □□	右の標識があるところでは，原動機付自転車は進入することができない。

Q28. T F ☐☐	Overtaking is prohibited near road bends.
Q29. T F ☐☐	The maximum statutory speed limit for a motorized bicycle towing a trailer is 20 kilometers per hour.
Q30. T F ☐☐	Pedestrian crossings and bicycle crossings differ only in the type of traffic they accommodate, but the method of passage for motorized bicycles is the same.
Q31. T F ☐☐	In muddy areas, it is advisable to reduce speed by using low gears and other means.
Q32. T F ☐☐	I drove the motorized bicycle at the speed indicated by the sign on the right. 
Q33. T F ☐☐	Even if there are lane markings extending into the intersection, it is not a priority road unless there is a priority road sign.
Q34. T F ☐☐	I overtook a motorized bicycle that was coasting near the top of a hill on an uphill slope.
Q35. T F ☐☐	Even for short durations, stopping within 10 meters before and after a railroad crossing is not allowed.
Q36. T F ☐☐	In the event of a major earthquake, evacuate on a maneuverable moped bicycle.
Q37. T F ☐☐	When turning the handlebars to the right while taking a curve, the motorcycle tends to lean to the left.
Q38. T F ☐☐	Motorized bicycles can travel on tram tracks at any time when trams are not in operation.
Q39. T F ☐☐	When a police officer is directing traffic with hand signals as shown in the illustration on the right, A and B have the same meaning. 
Q40. T F ☐☐	I sounded the horn to alert a child playing on the road and passed by them.
Q41. T F ☐☐	When traffic lights indicate green at an intersection on a two-lane road, motorized bicycles are allowed to make left turns or tight right turns.
Q42. T F ☐☐	Motorized bicycles are not allowed to enter areas where the sign on the right is displayed. 

問43 正誤 □□	荷物を積む場合は，方向指示器やナンバープレートなどがかくれないようにしなければならない。
問44 正誤 □□	盲導犬を連れた人が歩いているときは，一時停止か徐行してその人が安全に通れるようにしなければならない。
問45 正誤 □□	車は，前の車を追い越すためやむを得ないときには，軌道敷地内を通行することができる。
問46 正誤 □□	二輪車の点検をするとき，タイヤの空気圧は適正かどうかも点検する。

問47 時速30キロメートルで進行しています。この場合，どのようなことに注意して運転しますか？

(1) 正誤 □□ 子どもたちは，予測できない行動をとることがあるので，警音器を鳴らしてそのままの速度で進行する。

(2) 正誤 □□ 左側の子どもたちは道路上で遊んでいるため，急に車の前に出てくることはないので，このまま進行する。

(3) 正誤 □□ 右の路地の子どもは，急に車道に飛び出してくると思われるので，このままの速度で車道の左側端に寄って進行する。

問48 前車に続いて止まりました。坂道の踏切を通過するとき，どのようなことに注意して運転しますか？

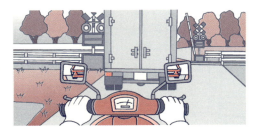

(1) 正誤 □□ 上り坂の発進は難しいので，発進したら前車に続いて踏切を通過する。

(2) 正誤 □□ 後続車がいるので渋滞しないように，前車のすぐ後ろについて進行する。

(3) 正誤 □□ 前車が発進しても，その先ですぐ停止してしまい，自分の車の入る余地がないかもしれないので，入れる余地があるか確認してから発進する。

Q43. T F ☐☐	When loading cargo, ensure that the turn signals and license plate are not obscured.
Q44. T F ☐☐	When a person with a guide dog is walking, stop or reduce speed to ensure their safe passage.
Q45. T F ☐☐	Vehicles may travel on streetcar tracks when necessary to overtake the preceding vehicle.
Q46. T F ☐☐	When inspecting a motorcycle, also check if the tire pressure is appropriate.

Q47. You are driving at 30 kilometers per hour. In this case, what should you pay attention to?

(1) T F ☐☐ Children may exhibit unpredictable behavior, so it's advisable to sound the horn and continue at the current speed.

(2) T F ☐☐ Children on the left side are playing on the road, so they are unlikely to suddenly come out in front of the car, thus it is safe to continue driving.

(3) T F ☐☐ Children in the alley on the right might suddenly dart onto the roadway, so it's advisable to continue at the current speed while keeping to the left side of the road.

Q48. When crossing a railroad crossing on a slope following the preceding vehicle, what should you pay attention to while driving?

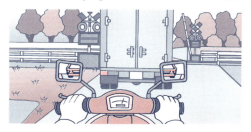

(1) T F ☐☐ Starting on upslope can be challenging, so once you've started, follow the preceding vehicle through the railroad crossing.

(2) T F ☐☐ Make sure not to cause congestion since there are trailing vehicles by closely following the preceding vehicle.

(3) T F ☐☐ If the preceding vehicle starts but immediately stops ahead, check if there's enough space for your vehicle to enter before starting.

第7回 解答と解説	間違えたら赤シートを当てて，覚えておきたいポイントを再チェックしよう！

◆・・・ひっかけ問題　　★・・・重要な問題

問1 誤	車両通行帯がない道路では，道路の**左側**に寄って通行しなければならない。
問2 ◆正	二輪車の点検では，ブレーキレバー，ブレーキペダル，チェーンの遊びは約**20〜30ミリメートル**が適当である。
問3 ★正	追越しのための右側部分はみ出し通行禁止の標示があるところでは，右側部分にはみ出さなければ**追越し**できる。
問4 誤	原動機付自転車の法定最高速度は**30キロメートル**毎時である。
問5 誤	免許証を交付されてからでないと，無免許運転になる。
問6 正	車輪のガタは，後輪よりも前輪のほうが運転に影響を与える。
問7 誤	バックミラーだけではなく，直接目視をして，安全確認のあとにゆるやかに車の流れに進入する。
問8 ★誤	マフラーの破損は，騒音公害の原因になり周囲に迷惑をかけるので，修理した後に運転する。
問9 正	この標識は「**車両通行止め**」の本標識と「**原付を除く**」の補助標識である。
問10 正	原動機付自転車は二段階右折の場合を除き，**直進，左折，右折**することができる。
問11 ◆誤	エンジンを**切って**押して歩いている場合のみ，歩行者としてみなされる。
問12 誤	この標識は「**車両通行区分**」を表していて，この通行帯は自動二輪車，原動機付自転車，自転車などの軽車両が通行できる。
問13 誤	環状交差点では入るときに合図はいらない。

226　第7回模擬試験　解答と解説

The Seventh Practice Test Answers and Explanations	If you make a mistake, use a red sheet to cover it and re-check the points you want to remember!

◆ · · · Tricky question ★ · · · Important question

Q1. F	On roads without vehicle lane designations, you must travel near **the left side** of the road.
Q2. ◆ T	During motorcycle inspections, it is appropriate for brake levers, brake pedals, and chain play to have a slack of about **20 to 30 millimeters**.
Q3. ★ T	In areas where there are signs prohibiting vehicles from protruding into the right side during **overtaking**, overtaking is permitted without protruding into the right side.
Q4. F	The maximum statutory speed limit for motorized bicycles is **30 kilometers** per hour.
Q5. F	It becomes driving without a license until the license is issued.
Q6. T	Wheel play affects driving more on the front wheels than on the rear wheels.
Q7. F	In addition to using rearview mirrors, visually confirm safety and smoothly enter the flow of traffic.
Q8. ★ F	Since muffler damage can cause noise pollution and inconvenience to others, drive only after repairs.
Q9. T	This sign indicates "**No Entry for Vehicles**" as the main sign and "**Except Motorized Bicycles**" as the supplementary sign.
Q10. T	Motorized bicycles can travel **straight**, **turn left**, or **turn right** except in the case of two-step right turns.
Q11. ◆ F	Only when the engine is **turned off** and the vehicle is pushed while walking will it be considered as pedestrian.
Q12. F	This sign represents "**Vehicle Lane Designation**," and vehicles such as motorcycles with engines and bicycles can use this lane.
Q13. F	No signaling is required when entering a roundabout.

The Seventh Practice Test Answers and Explanations 227

問14 正	みだりに進路変更すると，自分も危険であり，事故にもつながる。
問15 ◆誤	一時停止か徐行し，安全に通れるようにしなければならない。
問16 ★誤	原動機付自転車で交差点を右折するときに，車両通行帯が片側に３つ以上ある場合で信号があるところや，二段階右折の標識がある場合は，二段階右折しなければならない。
問17 ◆誤	補助標識により除外されていない一方通行の道路では，逆方向へ進行することはできない。
問18 ★誤	２本の白線で標示されている路側帯は，歩行者専用路側帯なので，車は中に入って駐停車することはできない。
問19 誤	追い越した車との間に安全な間隔をとって，前方に入る。
問20 正	環状交差点での転回を除いて，転回の合図は右折の合図と同じである。
問21 正	この標識は「横風注意」を表していて，減速するなど注意して運転する。
問22 誤	こう配の急な坂は，荷物の積み下ろしであっても駐停車禁止である。
問23 誤	原動機付自転車，軽車両，小型特殊自動車は，この場合左側に寄り進路を譲ればよい。
問24 ★誤	工事中でも，右側部分のはみ出しは最低限度にしてできるだけ左側部分を通行する。
問25 正	踏切内で動きがとれなくなるおそれがある場合，踏切に入ってはいけない。
問26 正	この標示は「車両通行区分」を示していて，原動機付自転車は左側の通行帯を通行する。
問27 ★誤	進路変更が終われば，速やかに合図をやめる。

228　第７回模擬試験　解答と解説

Q14. T	Making arbitrary lane changes is not only dangerous for oneself but also can lead to accidents.
Q15. ◆ F	One must come to **a stop** or **reduce speed** and ensure safe passage.
Q16. ★ F	When making a right turn at an intersection on a motorized bicycles, if there are three or more lanes on one side or there are signals or signs indicating **two-step right turns**, a two-step right turn must be made.
Q17. ◆ F	On one-way roads not exempted by supplementary signs, travel in **the opposite direction** is not allowed.
Q18. ★ F	Side strips marked with two white lines are **designated for pedestrians only**, and vehicles cannot park or stop inside.
Q19. F	Maintain a safe distance from overtaken vehicles before merging ahead.
Q20. T	Except for turning at a roundabout, the signal for turning is the same as that for **a right turn**.
Q21. T	This sign indicates "**Beware of crosswinds**" and drivers should reduce speed and drive cautiously.
Q22. F	**Parking or stopping is prohibited** even for unloading cargo on steep inclines.
Q23. F	For motorcycles with engines, light road vehicles, and special light equipments, it is sufficient to yield by moving to **the left side** in this scenario.
Q24. ★ F	Even during construction, protruding into the right lane should be minimized, and vehicles should predominantly use **the left lane**.
Q25. T	Do not enter a railroad crossing if there is a risk of being unable to move once inside.
Q26. T	This sign indicates "**Vehicle Lane Disignation**", with motorcycles with engines using the left lane for travel.
Q27. ★ F	Once a lane change is completed, promptly cease signaling.

The Seventh Practice Test Answers and Explanations 229

問28 正	道路の曲がり角付近は追越し禁止である。
問29 誤	原動機付自転車がリヤカーをけん引するときの法定最高速度は，**25 キロメートル**毎時である。
問30 正	横断歩道は歩行者，自転車横断帯は自転車が横断する場所のため，原動機付自転車や自動車の通行方法は同じである。
問31 正	ぬかるみのある場所では，低速ギアなど使い減速して，バランスをとりながら走行する。
問32 ★誤	標識にかかわらず，原動機付自転車は法定最高速度である**30 キロメートル**毎時を越えて運転してはいけない。
問33 ★誤	優先道路の標識がなくても，交差点の中まで**車両通行帯**の線が引かれている道路は，それだけで**優先道路**になる。
問34 誤	上り坂の頂上付近は，**追越し禁止**で徐行すべき場所のため，前車の後ろについて徐行し，追い越ししてはならない。
問35 正	踏切とその端から前後**10 メートル**以内の場所は**停車禁止**。
問36 ★誤	大地震で避難するときは，自動車や原動機付自転車をなるべく使用しない。
問37 正	カーブを走行中，ハンドルを右に切ると，反対の左に倒れようとする。
問38 誤	右左折や横断，転回などで横切るときや，標識で通行が認められている車など以外は，通行できない。
問39 正	イラストのように，同じ意味である。
問40 誤	警音器は鳴らさず，子どもの手前で**一時停止**や**徐行**で安全に進行する。
問41 ◆正	片側３車線以上の道路の交差点や標識によって二段階右折が指定されている交差点以外は**小回り右折**できる。

Q28. T	It is prohibited to overtake near road bends.
Q29. F	The maximum statutory speed limit for a motorized bicycle towing a trailer is **25 kilometers** per hour.
Q30. T	Crosswalks are for pedestrians, and bicycle crossings are for bicycles, so the traffic rules for motorized bicycles and cars are the same.
Q31. T	In muddy areas, reduce speed using low gears to maintain balance while riding.
Q32. ★ F	Regardless of signs, motorized bicycles must not exceed the statutory maximum speed limit of **30 kilometers** per hour.
Q33. ★ F	Even without priority road signs, roads where **the vehicle lane** markings extend into the intersection become **priority roads**.
Q34. F	Near the summit of an uphill stretch where passing is prohibited, one should reduce speed behind the preceding vehicle and **refrain from overtaking**.
Q35. T	**Stopping is prohibited** within **10 meters** before and after a railroad crossing.
Q36. ★ F	During major earthquakes, it is advisable to minimize the use of cars and motorized bicycles when evacuating.
Q37. T	When taking a curve, turning the handlebars to the right causes the vehicle to lean leftward.
Q38. F	You cannot proceed except for right or left turns, crossings, and turns where passage is permitted by signs.
Q39. T	As illustrated, they have the same meaning.
Q40. F	Instead of sounding the horn, safely proceed by **stopping** or **reducing speed** before reaching children.
Q41. ◆ T	Except for intersections on roads with three or more lanes on one side or at intersections designated for two-stage right turns by signs, **a tight right turn** is allowed.

The Seventh Practice Test Answers and Explanations   231

問42 正	「**車両進入禁止**」の標識のため，原動機付自転車は進入できない。
問43 正	方向指示器やナンバープレートは隠れないようにする。
問44 正	身体に障害がある歩行者が歩いている場合は，**一時停止**や**徐行**して安全に通れるようにしなければならない。
問45 ◆誤	「**軌道敷内通行可**」の標識によって認められた自動車が通行したり右折する場合を除いて，軌道敷内を通行することはできない。
問46 正	タイヤがすり減っていないか，空気圧は適正かなど点検する。

問47　道路上の**子どもたちの動き**に要注意！！
子どもは**予期しない動き**をするので注意して，速度を落として通行しよう。

(1)　**誤**　警音器は鳴らさず，**減速**して進行する。
(2)　**誤**　左側の子どもたちが**車道**に出てくるかもしれない。
(3)　**誤**　**減速**しないと左側の子どもが車道に出てきたときに，衝突して事故につながるかもしれない。

問48　坂道と，その先が**見えないこと**に要注意！！
坂道のため，前車が発進時に**後退**してくるかもしれません。また，踏切の先に自車が入れる**余地**があるか確認して発進しよう。

(1)　**誤**　踏切の直前で必ず，**一時停止**して安全確認する。
(2)　**誤**　前車が発進するときに，**後退**して衝突するかもしれないので間隔をあける。
(3)　**正**　自車が入る**余地**を確認してから，発進する。

Q42. T	Motorized bicycles cannot enter due to the "**No Entry for Vehicles**" sign.
Q43. T	Ensure that direction indicators and license plates are not hidden.
Q44. T	When physically impaired pedestrians are walking, **one must stop** or **reduce speed** to allow safe passage.
Q45. ◆ F	Except for when cars permitted by the "**Railroad Crossing Permitted**" sign pass or turn right, passage through the railroad crossing area is not allowed.
Q46. T	Inspect tires to ensure they are not worn out and check the tire pressure for adequacy.

Q47.   Be cautious of **children's movements** on the road!
Children may make **unexpected moves**, so reduce speed and proceed with caution.

(1)   F   Proceed without sounding the horn, **reducing speed** instead.
(2)   F   Children on the left side may come onto **the roadway**.
(3)   F   Failure to **decelerate** could lead to a collision if a child emerges onto the roadway from the left.

Q48.   Be cautious of slopes and **obscured views!**
Due to the slope, the preceding vehicle might **reverse** when starting. Also, ensure there is **space** for your vehicle beyond the railroad crossing before proceeding.

(1)   F   Always come to **a stop** and perform safety checks just before the railroad crossing.
(2)   F   Maintain a safe distance as the preceding vehicle might **reverse** when starting, potentially leading to a collision.
(3)   T   Ensure there is enough **space** for your vehicle to enter before proceeding with the start.

第8回 模擬試験	問1〜問46までは各1点，問47，48は全て正解して各2点。 制限時間30分，50点中45点以上で合格

●次の問題で正しいものは「正」，誤りのものは「誤」の枠をぬりつぶして答えなさい。

問1 正誤 □□	警察官や交通巡視員が，交差点以外の道路で手信号をしているときの停止位置は，その警察官や交通巡視員の3メートル手前である。
問2 正誤 □□	安全地帯のない停留所に，路面電車が停止しているときで乗降客がいない場合には，路面電車との間隔を1.5メートルあければ徐行して通行できる。
問3 正誤 □□	違法駐車をしていて放置車両確認標章を取りつけられたとき，その車を運転するときは取り除くことができる。
問4 正誤 □□	一方通行となっている道路で右折するときは，あらかじめ手前から道路の中央に寄り，交差点の中心の内側を徐行しなければならない。（環状交差点を除く）
問5 正誤 □□	バスの運行時間後，バスの停留所から10メートル以内に車を止めて，買い物に行った。
問6 正誤 □□	トンネルに入ると明るさが急に変わり，視力が急激に低下するので，入る前に速度を落とすようにする。
問7 正誤 □□	原動機付自転車で故障した原動機付自転車をロープでけん引するときは，ロープの真ん中に赤い布をつけなければならない。
問8 正誤 □□	重い荷物を積むとブレーキがよくきく。
問9 正誤 □□	右の標識は，本標識が表示する交通規制の終わりを意味している。
問10 正誤 □□	原動機付自転車は交通量が少ないときには自転車道を通行してもよい。
問11 正誤 □□	車を運転中，後方から緊急自動車が接近してきたが，交差点付近ではなかったので，徐行してそのまま進行を続けた。
問12 正誤 □□	右の標識のある場所は「右折禁止」を表している。
問13 正誤 □□	交差点付近の横断歩道のない道路を歩行者が横断していたが，車のほうに優先権があるので，横断を中止させて通過した。
問14 正誤 □□	信号機が赤色の灯火の信号でも，青色の灯火の矢印が左向きに表示されているときは，すべての車が左折することができる。

234　第8回模擬試験　問題

Eighth Practice Test.	Questions 1 to 46 are worth 1point each, while questions 47 and 48 are worth 2 points each if all answers are correct. The time limit is 30 minutes. To pass, you need 45 points out of 50.

● Fill in the box marked 'T' for the correct answers and 'F' for the incorrect ones in the following questions.

Q1. T F ☐☐	Police officers or traffic warden signal from 3 meters before the designated stopping position on roads other than intersections.
Q2. T F ☐☐	When a streetcar stops at a bus stop without a safety zone and there are no passengers boarding or alighting, vehicles can proceed cautiously by keeping a distance of 1.5 meters from the streetcar.
Q3. T F ☐☐	If a vehicle with an abandoned vehicle confirmation mark for illegal parking is found, the mark can be removed when driving the vehicle.
Q4. T F ☐☐	When making a right turn on a one-way street, drivers must first move towards the center of the road and proceed cautiously along the inner side of the intersection from the approach (excluding roundabouts).
Q5. T F ☐☐	After the operating hours of a bus, it's allowed to park a car within 10 meters of the bus stop for shopping purposes.
Q6. T F ☐☐	Entering a tunnel causes a sudden change in brightness and rapid deterioration in visibility, so drivers should reduce speed before entering.
Q7. T F ☐☐	When towing a broken down motor-assisted bicycle with another, a red cloth must be attached to the middle of the rope.
Q8. T F ☐☐	Brakes work better when carrying heavy loads.
Q9. T F ☐☐	The sign shown on the right indicates the end of the traffic regulation. 
Q10. T F ☐☐	Motorized bicycles are allowed to use bicycle lanes when traffic is light.
Q11. T F ☐☐	While driving, if an emergency vehicle approaches from behind but it's not near an intersection, reduce speed and continue driving.
Q12. T F ☐☐	The sign shown on the right indicates "No right turn". 
Q13. T F ☐☐	Pedestrians were crossing a road without a pedestrian crossing near an intersection, but vehicles have the right of way, so they were signaled to stop and allowed to pass.
Q14. T F ☐☐	Even when the traffic signal shows a red light, if a green arrow pointing left is displayed, all vehicles can make a left turn.

Eighth Practice Test. 235

問15 正誤 ☐☐	停留所で止まっている路線バスに追いついたときは，路線バスが発進するまで後方で一時停止していなければならない。
問16 正誤 ☐☐	車両通行帯のない道路では，中央線の左側ならばどの部分を通行してもよい。
問17 正誤 ☐☐	信号機の信号は横の信号が赤色であっても，前方の信号が青色であるとは限らない。
問18 正誤 ☐☐	横断歩道や自転車横断帯とその手前30メートル以内の場所では，追越しは禁止されているが，追抜きは禁止されていない。
問19 正誤 ☐☐	道路は公共の場所なので，交通の少ない道路ならば車庫代わりに使用してもよい。
問20 正誤 ☐☐	道路に平行して駐車している車の右側に並んで駐車することはできないが，停車はできる。
問21 正誤 ☐☐	原動機付自転車は右の標識のある交差点で右折するときは，交差点の中心のすぐ内側を徐行しなければならない。
問22 正誤 ☐☐	万一の場合に備えて，自動車保険に加入したり，応救護処置に必要な知識を身につけておく。
問23 正誤 ☐☐	原動機付自転車が普通自動車を追い越そうとするときは，その左側を通行しなければならない。
問24 正誤 ☐☐	原動機付自転車には，30キログラムまでの荷物を積むことができる。
問25 正誤 ☐☐	原動機付自転車の積み荷の制限は，ハンドルの幅いっぱいまでである。
問26 正誤 ☐☐	原動機付自転車は前方の信号が赤色であっても右のように青色の矢印が表示されているときは，すべての交差点で右折できる。
問27 正誤 ☐☐	進路の前方に障害物があるときは，あらかじめ一時停止か減速をして反対方向からくる車に道を譲らなければならない。
問28 正誤 ☐☐	初心運転者期間中に違反を犯し，初心運転者講習を受けなかったときは，免許が取り消される。
問29 正誤 ☐☐	原動機付自転車が，見通しのきく道路の曲がり角付近で，徐行している小型特殊自動車を追い越した。

Q15. T F ☐☐	When catching up with a stopped route bus at a bus stop, you must temporarily stop behind until the route bus starts.
Q16. T F ☐☐	On roads without vehicle lanes, you can travel on any part to the left of the centerline.
Q17. T F ☐☐	Even if the signal of the lateral traffic light is red, it doesn't necessarily mean that the signal ahead is green.
Q18. T F ☐☐	Overtaking is prohibited within 30 meters before pedestrian crossings, bicycle crossings, or their preceding areas, but passing is allowed.
Q19. T F ☐☐	Since roads are public places, it's permitted to use less trafficked roads as parking spaces.
Q20. T F ☐☐	While parking parallel to cars parked on the roadside, you cannot park alongside their right side, but stopping is allowed.
Q21. T F ☐☐	When a motorized bicycle turns right at an intersection with the sign shown on the right, it must proceed cautiously along the inner side of the intersection. 
Q22. T F ☐☐	To prepare for emergencies, it's advisable to have automobile insurance and acquire knowledge of first aid procedures.
Q23. T F ☐☐	When a motorized bicycle overtakes a standard motor vehicle, it must pass on the left side.
Q24. T F ☐☐	A motorized bicycle can carry loads up to 30 kilograms.
Q25. T F ☐☐	The limit for cargo on a motorized bicycle extends to the width of the handlebars.
Q26. T F ☐☐	Even if the signal ahead is red, if a green arrow is displayed as shown on the right, a motorized bicycle can turn right at all intersections. 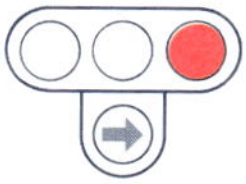
Q27. T F ☐☐	When there's an obstacle ahead in the path, you should come to a stop or reduce speed in advance and yield to oncoming traffic.
Q28. T F ☐☐	If a violation is committed during the bigginer driver period and the bigginer driver training is not completed, the license will be revoked.
Q29. T F ☐☐	A motorized bicycle overtook a special light eqipment that was driving at reduced speed near a curved corner on a road with good visibility.

Eighth Practice Test.

問30 正誤 ☐☐	横断歩道とその端から前後5メートル以内の場所は，駐車も停車もできない。
問31 正誤 ☐☐	エンジンブレーキを下り坂以外の場所で活用しても，制動距離には関係がない。
問32 正誤 ☐☐	右の路側帯の標示のある道路では，路側帯の幅が0.75メートルを超えるときだけ，その中に入って駐停車することができる。
問33 正誤 ☐☐	徐行や停止をする場合は，その行為をしようとするときに，手でも合図をすることができる。
問34 正誤 ☐☐	危険を防止するためやむを得ないときを除き，急ブレーキをかけるような運転をしてはならない。
問35 正誤 ☐☐	交通整理をしている警察官が灯火を横に振っているとき，その振られている灯火の方向へ進行するすべての車は，直進し，左折し，右折できる。
問36 正誤 ☐☐	遠心力は，カーブの半径が小さいほど，大きくなる。
問37 正誤 ☐☐	雨の日は，路面がすべりやすく停止距離も長くなるので，晴天のときよりも車間距離を多くとるのがよい。
問38 正誤 ☐☐	原動機付自転車で右左折の合図をする場合は，方向指示器によって行うだけでよく，手による合図は行ってはならない。
問39 正誤 ☐☐	右の標識のある場所では，駐停車が禁止されている場所であっても停車することができる。
問40 正誤 ☐☐	深い水たまりを通ると，ブレーキドラムに水が入りブレーキがきかなくなることがある。
問41 正誤 ☐☐	交通量が少ないときは，車両通行帯が黄色の線で区画されていても，いつでも進路を変えることができる。
問42 正誤 ☐☐	右の標示のあるところに歩行者がいる場合は，原動機付自転車は徐行して通行しなければならない。
問43 正誤 ☐☐	原動機付自転車でブレーキをかけるときは，ぬれた路面では後輪ブレーキをやや強くかける。

Q30. T F ☐☐	Crosswalks and areas within 5meters before and after them are prohibited for both parking and stopping.
Q31. T F ☐☐	Using the engine brake in places other than downhill does not affect the braking distance.
Q32. T F ☐☐	On roads with side strips shown on the right, it's only permitted to park in the lane if the width exceeds 0.75 meters. 
Q33. T F ☐☐	When intending to reduce speed or stop, hand signals can also be used in addition to other methods.
Q34. T F ☐☐	Unless necessary to prevent danger, drivers should avoid sudden braking.
Q35. T F ☐☐	When a traffic officer is waving a light horizontally, all vehicles can proceed in the direction indicated by the light for straight, left, and right turns.
Q36. T F ☐☐	Centrifugal force increases as the radius of a curve decreases.
Q37. T F ☐☐	On rainy days, roads become slippery and stopping distances increase, so maintaining a larger following distance than on sunny days is advisable.
Q38. T F ☐☐	When signaling for right or left turns on a motorized bicycle, it's sufficient to use the turn signal, and hand signals should not be used.
Q39. T F ☐☐	Even in areas where parking is prohibited according to this right-side sign, stopping is allowed. 
Q40. T F ☐☐	When passing through deep puddles, water may enter the brake drums, causing the brakes to fail.
Q41. T F ☐☐	Even on roads with yellow-marked vehicle lanes and light traffic, lane changes can be made at any time.
Q42. T F ☐☐	When pedestrians are present in areas indicated with this marking, motorized bicycles must proceed cautiously. 
Q43. T F ☐☐	When applying brakes on a motorized bicycle, slightly stronger pressure should be applied to the rear brake on wet roads.

Eighth Practice Test. 239

問44 正誤 ☐☐	一時停止の標識があるときは，停止線の直前で一時停止をして，交差する道路を通行する車などの進行を妨げてはいけない。
問45 正誤 ☐☐	発進の合図さえすれば，前後左右の安全を確認する必要はない。
問46 正誤 ☐☐	踏切を通過しようとしたとき，遮断機が降りはじめていたが，電車はまだ見えなかったので，急いで通過した。

問47　時速30キロメートルで進行しています。この場合どのようなことに注意して運転しますか？

(1) 正誤 ☐☐	バスのかげから歩行者が飛び出してくるかもしれないので，速度を落として走行する。
(2) 正誤 ☐☐	左側の歩行者のそばを通るときは，水をはねないように速度を落として進行する。
(3) 正誤 ☐☐	左側の歩行者は，車に気づかずバスに乗るため急に横断するかもしれないので，後ろの車に追突されないようブレーキを数回かけ，すぐに止まれるよう速度を落として進行する。

問48　時速30キロメートルで進行しています。カーブの中に障害物があるときは，どのようなことに注意して運転しますか？

(1) 正誤 ☐☐	前方のカーブは見通しが悪く，対向車がいつ来るか分からないので，カーブの入り口付近で警音器を鳴らし，自車の存在を知らせてから注意して進行する。
(2) 正誤 ☐☐	カーブ内は対向車と行き違うのに十分な幅がないので，対向車が来ないうちに通過する。
(3) 正誤 ☐☐	カーブの向こう側から対向車が自分の進路の前に出てくることがあるので，できるだけ左に寄って注意しながら進行する。

Q44. T F ☐☐	Stop at the stop line when there is a stop sign and do not obstruct the passage of vehicles crossing the intersection.
Q45. T F ☐☐	You don't need to confirm the safety of the front, rear, left, and right as long as you signal to start.
Q46. T F ☐☐	When trying to pass through a railroad crossing and the barrier starts to descend, but the train is not yet visible, I hurried to pass.

Q47. You are traveling at a speed of 30 kilometers per hour. What should you pay attention to while driving in this case?

(1) T F ☐☐ Pedestrians may jump out from behind the bus, so drive at a reduced speed.

(2) T F ☐☐ When passing pedestrians on the left side, reduce speed to avoid splashing water.

(3) T F ☐☐ Left-side pedestrians might suddenly cross without noticing the car to catch a bus, so apply the brakes a few times to avoid rear-end collisions and reduce speed to stop quickly.

Q48. You are driving at a speed of 30 kilometers per hour. What should you be careful of when there is an obstacle in the curve?

(1) T F ☐☐ The curve ahead has poor visibility, and it's uncertain when oncoming vehicles might come. Therefore, near the entrance of the curve, you will sound the horn to alert others of my presence and proceed with caution.

(2) T F ☐☐ Since there isn't enough width to pass oncoming vehicles safely in the curve, you will proceed through before any oncoming vehicles arrive.

(3) T F ☐☐ Since there's a possibility of oncoming vehicles appearing in my path from around the bend, you will proceed with caution, staying as far to the left as possible.

Eighth Practice Test.

第8回 解答と解説	間違えたら赤シートを当てて，覚えておきたいポイントを再チェックしよう！

◆・・・ひっかけ問題　　★・・・重要な問題

問1 ★誤	交差点以外で，横断歩道や自転車横断帯も踏切もないところで警察官や交通巡視員が手信号や灯火による信号をしているときの停止位置は，その警察官や交通巡視員の1メートル手前である。
問2 ★正	安全地帯のない停留所に，路面電車が停止しているときで乗降客がいない場合，路面電車との間隔を1.5メートルあけて徐行できる。
問3 正	放置車両確認標章を取りつけられた車を運転するときは，取り除くことができる。
問4 ◆誤	一方通行の道路を右折する場合は，道路の右側に寄り，交差点の中心の内側を徐行する。
問5 正	運行時間中に限り，バス，路面電車の停留所の標示板（標示柱）から10メートル以内の場所は，駐停車禁止である。
問6 ★正	トンネルを出入りするときは，減速する。
問7 誤	故障車をロープでけん引するときは，ロープの真ん中に0.3メートル平方以上の白い布をつける。
問8 ◆誤	重い荷物を積むと動く力が大きくなるため，ブレーキをかける強さが同じ場合でもききが悪くなる。
問9 正	この標識は交通規制の終わりを表している。
問10 誤	交通量が少なくても，自転車道を原動機付自転車は通行できない。
問11 ◆誤	徐行の義務はないので，道路の左側に寄って進路をゆずる。
問12 誤	この標識は「車両横断禁止」なので，この標識がある道路では右方向への横断をしてはならない。

The Eighth Practice Test Answers and Explanations	If you make a mistake, use a red sheet to cover it and re-check the points you want to remember!

◆ · · · Tricky question   ★ · · · Important question

Q1. ★ F	Outside of intersections, when police officers or traffic warden are signaling with hand gestures or lights and there are no pedestrian crossings, bicycle crossings, or railroad crossings, **the stopping position** is **one meter** before the officer or inspector.
Q2. ★ T	If there is no safety zone at a streetcar stop and there are no passengers boarding or alighting, you may drive at reduced speed while maintaining a distance of **1. 5 meters** from the streetcar.
Q3. T	When driving a vehicle with a deserted vehicle confirmation sign attached, you are allowed to **remove** the sign.
Q4. ◆ F	When turning right on a one-way road, you should keep to **the right side** of the road and drive at **reduced speed** on the inner side of the intersection.
Q5. T	Within 10 meters of a bus or streetcar stop signboard (or pole), **parking and stopping are prohibited during operating hours**.
Q6. ★ T	When entering or exiting a tunnel, you should **decelerate**.
Q7. F	When towing a broken- down vehicle with a rope, **a white cloth** of at least **0. 3 square meters** should be attached to the middle of the rope.
Q8. ◆ F	Loading heavy cargo increases the moving force, so even with the same braking strength, the effectiveness may decrease.
Q9. T	This sign indicates **the end** of traffic regulations.
Q10. F	Even if traffic is light, motorized bicycles are **not allowed** on bicycle lanes.
Q11. ◆ F	There is no obligation to drive at reduced speed, so yield the right of way by keeping to **the left side** of the road.
Q12. F	This sign indicates " **No Right Turn Crossing Ahead**, " so you must not cross to the right on roads with this sign.

The Eighth Practice Test Answers and Explanations   243

問13 ★誤	歩行者が横断しているときは，横断歩道のない交差点であっても，その通行を妨げてはいけない。
問14 正	青色の灯火の矢印が左向きに表示されているときは，**車は左折**することができる。
問15 ★誤	路線バスが発進の合図をしているとき以外は，**安全確認**して通過することができる。
問16 誤	追越しなどやむを得ない場合のほかは，道路の**左側**に寄って通行する。
問17 ◆正	横の信号が赤であっても，**前方の信号**が青であるとは限らないので，前方の信号を見る。
問18 誤	横断歩道や自転車横断帯とその手前 **30 メートル**以内の場所は追越し，追い抜きは禁止されている。
問19 誤	車の所有者は，道路でない場所に車庫や駐車場を用意しておかなければならない。
問20 誤	道路に平行して駐停車している車と並んで**駐停車**してはならない。
問21 ★正	この標識のある交差点では，交差点の中心のすぐ**内側**を徐行する。
問22 正	事故にあった場合に備えて，自動車保険に加入したり応急救護処置に必要な知識を身つけておく。
問23 誤	普通自動車が右折するため道路の中央に寄って通行しているときを除き，**右側**を追い越さなければならない。
問24 正	原動機付自転車の積載物の重量制限は **30 キログラム**以下である。
問25 ★誤	二輪車の積み荷の幅の制限は，**積載装置**の幅＋左右 0.15 メートル以下である。
問26 ◆誤	原動機付自転車は，二段階右折すべき交差点では**小回り右折**をすることができないので，前方の信号が赤色の場合は，青色矢印が表示されていても**停止**しなければならない。

Q13. ★ F	When pedestrians are crossing, even at intersections without crosswalks, their passage must not be obstructed.
Q14. T	When a green arrow signal is displayed pointing left, **vehicles are** allowed to make **a left turn**.
Q15. ★ F	Except when a route bus is giving a departure signal, you may proceed after **confirming safety**.
Q16. F	Unless unavoidable, vehicles should travel along **the left side** of the road.
Q17. ◆ T	Even if the side signal is red, it does not necessarily mean that **the forward signal** is green, so always check the forward signal.
Q18. F	Overtaking and passing are prohibited within **30 meters** before pedestrian crossings and bicycle crossings.
Q19. F	Vehicle owners must provide a garage or parking space off the road.
Q20. F	Do not **park or stop** alongside cars parked parallel to the road.
Q21. ★ T	At intersections with this sign, you must drive at reduced speed on **the inner side** of the intersection.
Q22. T	In preparation for accidents, it is important to have automobile insurance and knowledge of emergency first aid.
Q23. F	Except when standard motor vehicles are turning right and traveling toward the center of the road, vehicles must overtake on **the right**.
Q24. T	The weight limit for the cargo of motorized bicycles is **30 kilograms** or less.
Q25. ★ F	The width limit for the cargo of two-wheeled vehicles is the width of **the loading device** plus 0.15 meters on either side.
Q26. ◆ F	Since motorized bicycles cannot make **small right turns** at intersections where two-stage right turns are required, even if a green arrow is displayed, you must stop when the forward signal is red.

The Eighth Practice Test Answers and Explanations   245

問27 ★正	障害物があるときは，**一時停止**か**減速**して反対方向からの車に道をゆずる。
問28 誤	初心運転者講習を受けなかったら，再試験が行われて再試験が不合格や，受けなかった場合に免許が取り消される。
問29 ◆誤	曲がり角付近は，見通しがきくきかないに関係なく，**追越し禁止**の場所である。
問30 ★正	横断歩道とその端から前後**5メートル**以内の場所は**駐停車禁止**。
問31 ◆誤	エンジンブレーキと前後輪のブレーキを併用して速度を落とすと，**制動距離**を短くすることができる。
問32 ★誤	この路側帯は「**駐停車禁止の路側帯**」なので，**駐停車**することができない。
問33 ◆正	腕を斜め下に伸ばすと，徐行や停止の合図になる。
問34 正	急ブレーキは車輪の回転が止まり，スリップする原因になり危険なため，ブレーキは数回にわけてかける。
問35 ◆誤	二段階右折の交差点の原動機付自転車と軽車両は，**右折**できない。
問36 正	車にかかる遠心力は，カーブの半径が小さいほど大きくなり，**速度の2乗**に比例して大きくなる。
問37 正	雨の日は路面が滑りやすくなっているため，十分に注意して慎重に運転する。
問38 誤	必要に応じて手による合図も行ってよい。
問39 正	この標識は「**停止可**」を表しているので，停車することができる。
問40 正	ブレーキドラムに水が入ると，ブレーキの**ききが悪くなったり，きかなくなったり**することがある。

246　第8回模擬試験　解答と解説

Q27. ★ T	When there is an obstacle, either come to **a stop** or **reduce speed** to yield to vehicles from the opposite direction.
Q28. F	If you don't take the novice driver training, you will be required to retake a test, and if unsuccessful or not taken, your license may be revoked.
Q29. ◆ F	Near corners, regardless of visibility, **overtaking is prohibited**.
Q30. ★ T	Within **5 meters** before and after a pedestrian crossing and its edges, **parking and stopping are prohibited**.
Q31. ◆ F	Using the engine brake and both front and rear brakes together can shorten **braking distance**.
Q32. ★ F	This roadside is designated as "**No Parking or Stopping**," **so parking or stopping** is not allowed.
Q33. ◆ T	Extending your arm diagonally downward serves as a signal for reducing speed or stopping.
Q34. T	Applying brakes abruptly can cause wheel rotation to stop and lead to slipping, making it dangerous. Therefore, brakes should be applied gradually over several instances.
Q35. ◆ F	Motorized bicycles and light road vehicles cannot make two-stage **right turns** at intersections.
Q36. T	Centrifugal force acting on a car increases as the radius of curvature decreases, and it is proportional to **the square of the speed**.
Q37. T	On rainy days, roads become slippery, so drive cautiously and with extra care.
Q38. F	Hand signals may be used as needed.
Q39. T	This sign indicates "**Stopping Allowed**," so you can stop.
Q40. T	When water enters the brake drum, it can **impair or completely inhibit** brake function.

The Eighth Practice Test Answers and Explanations **247**

問41 ◆誤	緊急自動車に進路をゆずる時や，**道路工事**を避けるときなど除いて，黄色の線の車両通行帯では，**進路**を変更することができない。
問42 正	この標示は「**安全地帯**」を表示しているので，歩行者がいる場合は徐行する。
問43 ◆正	二輪車のブレーキをかけるとき，乾燥した路面では前輪ブレーキを，ぬれた路面では後輪ブレーキをやや強くかけるとよい。
問44 正	**一時停止**の標識のあるところでは，停止線の直前で**一時停止**する。
問45 誤	発進の合図だけでなく，前後左右の安全を確認して，方向指示器などで発進合図を行う。
問46 ★誤	遮断機が降りはじめていたり，警報機が鳴っているときは，踏切に入ってはいけない。

問47　**歩行者の動きと水たまり**に要注意！！
歩行者が**道路を横断**したり，歩行者に**水がはねてしまったり**するかもしれないので，注意しよう。

(1)　正　バスのかげから歩行者が**飛び出してくる**かもしれないので，注意する。
(2)　正　歩行者に水がはねたら迷惑がかかるので，**速度を落として**通過する。
(3)　正　ブレーキを**数回にわけて**かけると，後続車への追突防止となります。

問48　**対向車**が見えないこと，**工事中**に要注意！！
対向車が見えず，接近しているかもしれないので注意しよう。また工事中なので，右側を通行してくるおそれもあるので，慎重に運転しよう。

(1)　正　**警音器**を鳴らすと，対向車に自車の**存在**を知らすことができる。
(2)　誤　対向車が来て，**衝突**するかもしれないので危険である。
(3)　正　対向車が来るおそれがあるので，できるだけ**左**に寄って進行する。

Q41. ◆ F	Except when yielding to **emergency vehicles** or avoiding **road construction**, you cannot change lanes in lanes marked with yellow lines.
Q42. T	This marking indicates a "**safe zone**," so you should proceed with caution if pedestrians are present.
Q43. ◆ T	When applying brakes on a motorcycle, it's advisable to apply the front brakes on dry surfaces and slightly stronger pressure on the rear brakes on wet surfaces.
Q44. T	At locations with **stop signs**, come to **a stop** just before the stop line.
Q45. F	In addition to signaling to start, ensure safety in all directions by checking before signaling with indicators.
Q46. ★ F	Do not enter a railroad crossing when the barriers are beginning to descend or when the alarm is sounding.

Q47.　　Be cautious of **pedestrians'movements** and **puddles!**
Be attentive as pedestrians may **cross the road** or get **splashed by water**, requiring your attention.

(1)　T　Be cautious as pedestrians may **emerge suddenly** from behind buses.
(2)　T　**Reduce speed** to avoid splashing pedestrians with water, which could cause inconvenience.
(3)　T　Applying the brakes in **intervals** helps prevent rear-end collisions with following vehicles.

Q48.　　Be cautious as **oncoming vehicles** may not be visible, and be aware of **ongoing roadworks**.
Since oncoming vehicles might be approaching without being seen, drive carefully. Additionally, due to the ongoing roadworks, there's a possibility of traffic flowing on the right side, so proceed with caution.

(1)　T　By sounding **the horn**, you can alert oncoming vehicles to your **presence**.
(2)　F　It's dangerous because there's a possibility of **a collision** with oncoming vehicles.
(3)　T　Since there's a possibility of oncoming vehicles, proceed as far to **the left** as possible.

The Eighth Practice Test Answers and Explanations　249

あとがき

すべては観察からはじまる

日常を観察することで、気づきを得ることができます。気づきだけではなく、仮説を立て検証する。これらを繰り返すことにより、観察は最速で最善の結果を招くように思います。

観察により、弊社は投資や複数の事業を構築するようになりました。また複数の発明や出版も世に問うことができました。

今回は、弊社取締役柳井正彦が数ある国家試験の中で唯一、教習所やセミナー、学校などが存在しない資格が存在することに気がつきました。

その資格とは原付き免許です。

原付き免許試験だけは、独学以外の方法では受験ができません。独学で受験しようとすると受験参考書が必要です。

書店にはこれらの需要に合わせるように受験参考書が並んでいます。同時に外国語で書かれた原付き免許受験参考書が少ないことも観察できました。

この観察により 2020 年に「ベトナム語で解説　原付き免許　めざせ一発合格」が出版されました。本書はこの英語版です。

ご縁があり、読者と知り合えたのですから、読者も自身の廻りを観察し、気づきがあれば仮説検証を繰り返し、世に問うてみてください。

きっと面白い物語がはじまりますよ。本を手にとってくださりありがとうございます。

株式会社知財事業研究所
代表取締役　大賀信幸

Afterword

All things start with observation.

By observing everyday life, we can gain new insights. Not only can we gain insights, but we can also formulate hypotheses and test them. By repeating this process, observation seems to bring about the fastest and best results.

Through observation, our company has been able to develop investments and multiple businesses. It has also allowed us to bring several inventions and publications to the world.

This time, our director Masahiko Yanai noticed that among the many national exams, there is only one qualification that lacks the existence of driving schools, seminars, or courses.

That qualification is the moped license.

The moped license exam is the only one that can only be taken through self-study. If you intend to prepare for the exam through self-study, you will need a study guide.

In bookstores, study guides are lined up to meet this demand. At the same time, it was observed that there are very few study guides for the moped license available in foreign languages.

This observation led to the publication of "Moped License: Aim for passing on the first try, Explained in Vietnamese" in 2020. This book is its English version.

To the readers, for this opportunity connected with our company, I encourage you to observe your surroundings, and if you gain any insights, continue to test your hypotheses and bring them to the world.

Surely, an interesting story will begin. Thank you for picking up this book.

Intellectual Property Business Laboratory
CEO Nobuyuki Oga

監修・著者（追加）

株式会社知財事業研究所
設立：2020 年 5 月
代表取締役：大賀信幸
取締役：柳井正彦
取締役：山本英彦（弁理士）

ミッション：「知の流通する社会」を実現する。

　「知の流通する社会」とは、一人が生み出した知的財産が、他者の知的財産の基礎となり、それが連鎖的に発展していくことで、より豊かな社会を築くというビジョンに基づいています。この理念の下、知的財産権（著作権、特許、商標など）とビジネスの融合に関する研究を行っています。

　具体的な活動として、自社コンテンツの書籍出版や、他者の書籍出版の支援、自社の特許や商標などの工業所有権の取得および事業への活用、他社の工業所有権の活用に関するコンサルティングなどを手掛けています。また、近年は知的財産と AI（人工知能）の関係に研究の焦点を当てており、急速に普及した生成 AI の事業活用およびそれに伴う知的財産の課題についても積極的に取り組んでいます。

　今回の著書の出版にあたっては、AI 技術を駆使した新しい出版の形を提案する試みとして、従来の出版プロセスに比べ、短期間で高品質なコンテンツを提供することを可能にしました。また、AI を活用して生まれた書籍における著作権の帰属についても考察する機会となりました。

　知的財産権をいかに世に問うか方策の方法のひとつとして、今回の出版に繋がりました。

代表取締役　　大賀信幸

発明者：
継手ジョイント
負圧の生じないポンプ
債務弁済システム

著作者：
１級建築士に面白ほど受かる本
他著書 20 冊以上

経営者：
株式会社ワークス・ワン
株式会社住まい壱番館
株式会社知財事業研究所　他

取締役　柳井正彦

株式会社ワークスワン　取締役
著作
営業の天才
ベトナム語で解説　原付き免許　めざせ一発

取締役　山本英彦（弁理士）

IPUSE 特許事務所 代表弁理士
中小製造業勤務
折り曲げラベル入り卵パックの特許取得と事業化

Intellectual Property Business Laboratory Co., Ltd.

Founded: May 2020
CEO: Nobuyuki Oga
Director: Masahiko Yanai
Director: Hidehiko Yamamoto (Patent Attorney)

Mission: "Creating a Society of Knowledge distribution "

A "society where knowledge distribution" is based on the vision that intellectual property created by one person becomes the foundation for others' intellectual property, leading to its continued development in a chain, ultimately building a more enriched society. Under this philosophy, we conduct research on the integration of intellectual property rights (such as copyrights, patents, and trademarks) with business.

Our specific activities include publishing our own content, supporting the publication of others' works, acquiring and utilizing our own patents and trademarks (industrial property rights), and providing consulting on the utilization of industrial property rights owned by other companies. Recently, our research has focused on the relationship between intellectual property and AI (artificial intelligence). We are actively addressing issues related to the business application of generative AI, which has rapidly gained popularity.

With the publication of this book, we aimed to propose a new form of publishing by leveraging AI technology, making it possible to provide high-quality content in a shorter time compared to traditional publishing processes. This endeavor also provided an opportunity to explore the ownership of copyrights in books generated with the help of AI.

This publication came about as one of the approaches to explore how intellectual property rights can be effectively brought to the public.

CEO: Nobuyuki Oga

Inventor:
- Coupling Joint
- Pump with No Negative Pressure
- Debt Repayment System

Author:
- "A Book to Easily Pass the First-Class Architect Exam"
  Over 20 other books

Executive:
- Works One Co., Ltd.
- Sumai Ichibankan Co., Ltd.
- Intellectual Property Business Laboratory Co., Ltd. and others

Director: Masahiko Yanai

Director at Works One Co., Ltd.
Author:
"Genius in Sales"
"Moped License: Aim for One-Time Success" (Explained in Vietnamese)

Director and Patent Attorney: Hidehiko Yamamoto

IPUSE Patent Office

Business Performance
- Acquired and commercialized the patent for a "Folded Label Egg Pack"

弊社ホームページでは，書籍に関する様々な情報（法改正や正誤表等）を随時更新しております。ご利用できる方はどうぞご覧下さい。http://www.kobunsha.org 正誤表がない場合，あるいはお気づきの箇所の掲載がない場合は，下記の要領にてお問い合せ下さい。

英語で解説　原付免許　めざせ一発合格

監　　　修	株式会社　知財事業研究所
編 著 者	大賀信幸・山本英彦・柳井正彦
印刷・製本	亜細亜印刷株式会社

発 行 所	株式会社 弘 文 社	〒546-0012 大阪市東住吉区 中野2丁目1番27号 ☎　（06）6797—7 4 4 1 FAX（06）6702—4 7 3 2 振替口座　00940—2—43630 東住吉郵便局私書箱1号
代 表 者	岡﨑　　靖	

ご注意
（1）本書は内容について万全を期して作成いたしましたが，万一ご不審な点や誤り，記載もれなどお気づきのことがありましたら，当社編集部まで書面にてお問い合わせください。その際は，具体的なお問い合わせ内容と，ご氏名，ご住所，お電話番号を明記の上，FAX，電子メール（henshu1@kobunsha.org）または郵送にてお送りください。
（2）本書の内容に関して適用した結果の影響については，上項にかかわらず責任を負いかねる場合がありますので予めご了承ください。
（3）落丁・乱丁本はお取り替えいたします。